Applications of Semiconductor Optical Amplifiers

Applications of Semiconductor Optical Amplifiers

Special Issue Editor

Kyriakos E. Zoiros

MDPI • Basel • Beijing • Wuhan • Barcelona • Belgrade

MDPI

Special Issue Editor
Kyriakos E. Zoiros
Democritus University of Thrace
Greece

Editorial Office
MDPI
St. Alban-Anlage 66
Basel, Switzerland

This is a reprint of articles from the Special Issue published online in the open access journal *Applied Sciences* (ISSN 2076-3417) from 2017 to 2018 (available at: http://www.mdpi.com/journal/applsci/special_issues/optical_amplifiers)

For citation purposes, cite each article independently as indicated on the article page online and as indicated below:

LastName, A.A.; LastName, B.B.; LastName, C.C. Article Title. *Journal Name* **Year**, *Article Number*, Page Range.

ISBN 978-3-03897-170-2 (Pbk)
ISBN 978-3-03897-171-9 (PDF)

Contents

About the Special Issue Editor

Kyriakos E. Zoiros (born 1973 in Thessaloniki, Greece) is an Associate Professor at the Department of Electrical and Computer Engineering, Democritus University of Thrace (DUTH), Xanthi, Greece. He holds a Ph.D. degree on optical communications from the Photonics Communications Research Laboratory, National Technical University of Athens, Greece. He is the author or coauthor of more than 100 journal and conference papers, some of which were invited, as well as of five book chapters. His published work has received a considerable number of citations (>1000, h-index = 20, source Scopus). In 2009, he was seconded to the Optical Communications Research Group, University of Limerick, Ireland, and in 2013 to the École Nationale d'Ingénieurs de Brest (ENIB), France. His current research interests include applications of semiconductor optical amplifiers, applications of microring resonators, microwaves photonics, and free space optical communications.

Preface to "Applications of Semiconductor Optical Amplifiers"

During recent years, the technology of semiconductor optical amplifiers (SOAs) has been evolving remarkably and has matured to the point where it is presently established as a key enabler for the development, implementation, optimization, and overall establishment of photonic circuits, subsystems, and networks. Owing to the outstanding advancements that have been achieved in the field, SOAs are commercially available devices that exhibit several important properties, such as strong nonlinearities, low power consumption, wavelength flexibility, a large dynamic range, fast response, broadband and versatile operation, small footprint, and the capacity for scalable integration in single chips at an affordable cost. These attractive characteristics have rendered SOAs core elements for the accomplishment of critical and indispensable tasks at the fundamental and system-oriented level. Thus, SOAs have widely been adopted both by the research community and the industrial sector as a principal technological platform for the realization of a diverse range of applications with high performance.

Given the huge practical potential of SOAs, this book contains papers published within the frame of a Special Issue on 'Applications of Semiconductor Optical Amplifiers', with a twofold aim. On one hand, it sought to address, present, and investigate modern applications of SOAs, and on the other hand, to explore and highlight trends, challenges, and perspectives to motivate efforts toward the continuous exploitation of these active modules in a feasible, innovative, and global manner. This book collates the Special Issue papers reporting on the significant results obtained from the cutting-edge research conducted by experts in the field. The compilation can provide useful knowledge and open new horizons regarding SOA-enabled applications, such as direct signal amplification, external modulation, all-optical signal processing, all-optical memories, photonic integrated circuits, photonic switching, optical code division multiple access systems, and passive optical networks.

The research field of integrated optical memories was covered by C. Vagionas, P. Maniotis, S. Pitris, A. Miliou, and N. Pleros. This paper describes a novel application scenario for optical memories based on monolithically integrated SOA and Mach-Zehnder interferometer layouts/arrangements. In this context, it proposes an alternative path to circumvent the mismatch between the rapidly growing optical transmission line rates and electronic processing speeds to facilitate technological evolution in the era of the Internet of Things.

The use of SOAs for optical signal amplification applications is reported in two papers.

Z.V. Rizou, K.E. Zoiros, and A. Hatziefremidis present a theoretical analysis and benchmarking of two basic optical notch filters employed to compensate for the SOA pattern effect. The comparison reveals the performance merits of each filter that allow it to favorably compete against the other to efficiently address the pattern-dependent operation of SOAs and assisting the latter in serving linear amplification applications with improved performance.

S.P. Ó Dúill, P. Landais, and L.P. Barry propose multi-section SOAs as pre-amplifiers for short range optical communication links within datacenters. By creating and employing a simplified multi-section SOA model, they evaluate the performance benefits offered by these special type SOAs, which are shown to exhibit a better input power dynamic range than conventional single-section SOAs when amplifying single- as well as multi-channel signals in the advanced modulation format (four-level pulse amplitude modulation/PAM4).

Reflective semiconductor optical amplifiers (RSOAs) are the subject of two papers. S.A. Gebrewold, R. Bonjour, R. Brenot, D. Hillerkuss, and J. Leuthold conduct a comparative study of the capacity increase brought in wavelength division multiplexing (WDM) passive optical network (PON) architectures by exploiting bit- and power-loading discrete multi-tone (DMT) modulation in upstream colorless transmitters using RSOAs. Three different RSOA-based schemes are compared against appropriate evaluation criteria whose measurement and analysis allows them to specify and highlight performance and cost trade-offs. Concurrently, record high line rates in both back-to-back and transmission experiments were obtained and reported for all three configurations.

Z.V. Rizou and K.E. Zoiros demonstrate the feasibility of using a single microring resonator (MRR) as an optical notch filter to enable RSOA direct modulation at an extended data rate compared to what is possible with the RSOA alone. To this aim, it was investigated and specified how the MRR should be designed to improve the encoded signal characteristics for RSOA direct modulation applications.

The exploitation of SOAs for photonic switching purposes is addressed in two papers.

N. Calabretta, W. Miao, K. Mekonnen, and K. Prifti present a novel photonic WDM optical cross-connect node based on SOAs that allows switching data signals in wavelength, space, and time to fully exploit statistical multiplexing. Also, they report on the experimental assessment of this core building block for interconnecting network elements as well as computing and storage resources, which verifies the advantages of using SOAs to realize the WDM cross-connect switch in terms of transparency, switching speed, photonic integrated amplification for lossless operation, and gain equalization.

R. Stabile reviews the current status of fast reconfigurable medium-scale indium phosphide (InP) integrated photonic switch matrices based on the use of SOA gates. The focus is on broadband and cross-connecting monolithic implementations that grant multi-input/output port and channel connectivity based on a packet-compliant SOA multi-stage switching matrix. The opportunities for increasing connectivity, enabling nanosecond-order reconfigurability, and introducing distributed optical power monitoring at the physical layer are highlighted. Furthermore, complementary architectures based on resonant switching elements developed on the same material platform for power efficient switching are considered. Lastly, performance projections related to the physical layer are presented, and strategies for improvements in view of opening a route towards large-scale power efficient fast reprogrammable photonic integrated switching circuits are discussed.

The leveraging of SOAs as nonlinear elements in the context of all-optical functionalities is treated by Y. Lin, A.P. Anthur, S. P. Ó Dúill, F. Liu, Y. Yu, and L.P. Barry. In particular, a wavelength converter is developed, which is comprised of an SOA that takes advantage of four wave mixing (FWM) and a fast-switching sampled grating distributed Bragg reflector (SG-DBR) tunable laser as one of the pump sources. By studying phase noise issues in FWM as well as vector theory in SOAs in conjunction with a detailed characterization of the SGDBR laser, rapid and reconfigurable wavelength conversion is experimentally demonstrated on bursty/packet data of advanced modulation formats towards dynamic, adaptive and bandwidth-efficient next generation transparent optical networks.

Finally, the research activity on optical code division multiple access (OCDMA) is presented by M.S. Ahmed and I. Glesk. This paper explores the use of SOAs in the transmitter side of such a system based on multi-wavelength picosecond code carriers for the mitigation of the temporal distortion of an OCDMA auto-correlation affected by the temperature-induced dispersion changes in a fiber optic transmission link. To this aim, it is shown both experimentally and using simulations, that a distorted OCDMA auto-correlation due to the temperature-induced fiber dispersion can be corrected by manipulating the chirp of code carriers when traversing a biased SOA prior to entering the transmission link.

I would like to thank all those who contributed to making this book possible.

Kyriakos E. Zoiros
Special Issue Editor

applied
sciences

MDPI

Article

Integrated Optical Content Addressable Memories (CAM) and Optical Random Access Memories (RAM) for Ultra-Fast Address Look-Up Operations

Christos Vagionas *, Pavlos Maniotis, Stelios Pitris, Amalia Miliou and Nikos Pleros

Department of Informatics, Aristotle University of Thessaloniki, 54124 Thessaloniki, Greece;
ppmaniot@csd.auth.gr (P.M.); skpitris@csd.auth.gr (S.P.); amiliou@csd.auth.gr (A.M.); npleros@csd.auth.gr (N.P.)
* Correspondence: chvagion@csd.auth.gr; Tel.: +30-231-0990-588

Academic Editor: Kyriakos E. Zoiros
Received: 30 May 2017; Accepted: 4 July 2017; Published: 7 July 2017

Abstract: Electronic Content Addressable Memories (CAM) implement Address Look-Up (AL) table functionalities of network routers; however, they typically operate in the MHz regime, turning AL into a critical network bottleneck. In this communication, we demonstrate the first steps towards developing optical CAM alternatives to enable a re-engineering of AL memories. Firstly, we report on the photonic integration of Semiconductor Optical Amplifier-Mach Zehnder Interferometer (SOA-MZI)-based optical Flip-Flop and Random Access Memories on a monolithic InP platform, capable of storing the binary prefix-address data-bits and the outgoing port information for next hop routing, respectively. Subsequently the first optical Binary CAM cell (B-CAM) is experimentally demonstrated, comprising an InP Flip-Flop and a SOA-MZI Exclusive OR (XOR) gate for fast search operations through an XOR-based bit comparison, yielding an error-free 10 Gb/s operation. This is later extended via physical layer simulations in an optical Ternary-CAM (T-CAM) cell and a 4-bit Matchline (ML) configuration, supporting a third state of the "logical X" value towards wildcard bits of network subnet masks. The proposed functional CAM and Random Access Memories (RAM) sub-circuits may facilitate light-based Address Look-Up tables supporting search operations at 10 Gb/s and beyond, paving the way towards minimizing the disparity with the frantic optical transmission linerates, and fast re-configurability through multiple simultaneous Wavelength Division Multiplexed (WDM) memory access requests.

Keywords: Optical Content Addressable Memories; Optical Random Access Memories; Address Look Up; Optical Matchline; Semiconductor Optical Amplifier Mach Zehnder Interferometers; photonic integration; monolithic InP platform

---

1. Introduction

The last decades have been marked by the widespread use of bandwidth-hungry internet applications by multiple wireless users and cloud-network devices always connected online. This has led to an immense Internet expansion [1], enabled by the rapid advances in photonic integration [2] and optical transceiver technologies that achieve doubling of the optical transmission line-rates every year [3]. Meanwhile, Internet topologies have strongly relied on resilient multi-homing techniques and on the Virtual Private Network (VPN) for enhanced network resiliency or security, necessitating additional physical or logical communication links [4]. The insatiable interconnectivity demands have resulted in an enormous surge in the number of addressable end-points [1,4], even running up to the complete exhaustion of the unallocated IPv4 address pool [5], enforcing the use of the next generation IPv6 protocol. IPv6 offers a higher availability for address space, but at the same time, quadruples the needs of Address Look-Up (AL), while scaling at a frantic annual growth rate of 90% [6]. As a result, the Default Free Zone (DFZ) has been constantly expanding, with the Routing Information Base (RIB) of Internet core routers increasing up to 700 K prefix-entries [7], requiring increasingly more search intensive operations to resolve the outgoing port of an incoming packet. Moreover, recent studies on the content centric nature of today's internet usage have even inspired Content Centric Networking (CCN) [8] that investigates a clean slate future internet, where packet forwarding will operate based on content addressing, instead of addressing the destination end-host, yet this would tremendously scale the respective name look-up requirements. As performance sensitive AL operations have to be performed at wire speed upon the arrival of the packet [4], this has rendered software algorithmic search approaches with sequential access schemes to Random Access Memories (RAMs) as rather impractical since the early 2000s [4], necessitating specialized electronic hardware AL-solutions [9,10].

Presently, routers rely on electronic Content Addressable Memories (CAMs) that facilitate AL table functionalities within one clock cycle [10]. CAMs offer content-based addressing of the stored data, instead of location-based addressing, forming an alternative to conventional RAMs of computing architectures [11,12]. Specifically, upon the arrival of a packet, its destination address is inserted into a CAM-table for a fast parallel comparison across the AL memory contents, and upon a match, the outgoing port is obtained for next hop routing. In order to comply with the Classless Interdomain Routing (CIDR) [13], CAMs have also been equipped with ternary features, to support wildcard bits of network subnets, that mask the stored RIB-prefixes at arbitrary bit-positions with a "logical X" value [10]. Early fast demonstrations of such Ternary CAM (T-CAM) devices built on 250 nm Complementary Metal–Oxide–Semiconductor (CMOS) nodes supported content comparisons at 260 MHz [14], while similar T-CAMs at 180 nm [15], 130 nm [16], or 62 nm [17] CMOS nodes achieved maximum frequencies of 210 MHz, 200 MHz, and 400 MHz, respectively. Despite the rich variety of optimization techniques of mature electronic technology, state of the art electronic CAMs are scaling at a slow growth rate [18,19], and even by shifting to advanced 28 nm CMOS Fully Depleted Silicon-On-Insulator (FD-SOI) [20], only footprint and power reductions have been achieved, with frequencies still lying around 370 MHz. These results imply that electronic T-CAMs are hard-limited by the underlying interconnect network and can rarely reach the barrier of 1 Gb/s. This barrier was only recently broken using alternative non-optimal techniques that may use early predict/late-correct schemes [21], which are yet known to be heavily dependent on data patterns [17]. A second speed enhancement technique suggests inserting four T-CAM arrays performing in parallel at a slower rate of 4 × 400 MHz [6], necessitating even more complex Application Specific Integrated Circuit (ASIC) for deserialization and further exacerbating the energy requirements of routers, which reached their rack-power density limits in 2005 [4,22,23]. To this end, electronic CAM speeds seem inefficient to keep up with the frantic optical linerates of 100 Gb/s and beyond [1]. This performance disparity has been placing an increasingly heavy load on the shoulders of electronic CAMs, enforcing energy-hungry, cost-expensive optoelectronic header conversions with subsequent data-rate down-conversion [22–24], in order to perform AL searches in the MHz-regime. Furthermore, the migration towards Software Defined Networks (SDN) and OpenFlow networks enforces a dynamic operation with frequent updates of network topologies and multiple real time changes in the RIB-list [25]. Internet routers are experiencing 100 updates per second and potentially reaching upwards of 1000 per second [26], during which AL operations are stalled and the router remains idle, considerably limiting the performance. This, in-turn, requires fast Write operations to the CAM-table, with short latencies [25–28], to enable fast re-configurability of the network and rapid updating of the AL table.

Having detected the impact of slow-performing AL operations, optics have tried to circumvent the associated delays, mainly through the use of optical labeling and header processing schemes. These insert a bit-serial label in front of the payload or various multiplexing schemes, to route the data based on this label instead of the actual destination IP address [29], yet they typically utilize lower label data-rates, hoping to retain compatibility with slow CAM speeds. At the same time, optical memories have undergone two decades of developments and are now on the verge of developing higher capacity [30], programmable [31], and/or non-volatile [32] devices towards more practical memory sub-systems. Initially, optical memories were conceived as high bandwidth alternatives of electronic RAMs to overcome the "Memory Wall", achieving multiple elementary Flip-Flops (FFs) with high speed and low power consumption credentials [33–41], including coupled SOAs [34], III-V-on-SOI microdisk lasers [35], and polarization bistable Vertical Cavity Surface Emitting Laser (VCSEL) [36], as well as coupled SOA-based Mach-Zehnder Interferometers (SOA-MZIs) [33,37]. Optical FFs, serving as optical storing units, were then combined with random access controlling gates, forming functional optical RAMs, so far demonstrated either as fully functional architectures using mature SOA-based devices [37–39] or as discrete components based on photonic crystals [40,41]. These have both been experimentally shown to support speeds beyond 10 Gb/s, while in-depth frequency theoretical memory speed analyses [42,43] and validated time-domain SOA-based memory simulations [43,44] have revealed potential rates of up to 40 Gb/s. Furthermore, by combining optical Column/Row Address Selectors [45,46] and optical Tag Comparators [47], the first designs of a complete optical cache memory architecture for high-performance computers revealed a 16 GHz operation via physical layer simulations [48]. All of these have increased the maturity of optical memories towards penetrating the computing domain, where the use of electronics is so far undisputable, whereas in optical networks, optical FFs have been suggested for contention resolution [49].

Following the paradigm of optical RAMs, optical alternatives of CAM architectures may facilitate similar advances and speed enhancements towards ultra-fast router AL memories in the high-end router domain. In this regime, some preliminary first steps, stemming from our group, have managed to develop the first photonic alternative CAM-based elements [50,51], which is the main focus of this paper. More specifically, in Section 2, we initially discuss the architectures and main functional building blocks of electronic AL memories, followed by the development of monolithic photonic integration for optical FFs and RAM memories on a monolithic InP platform [33,39] in Section 3. In Section 4, we present the first experimental proof-of-principle of an all-optical Binary CAM (B-CAM) cell architecture at 10 Gb/s [50]. This architecture is later extended in a more advanced all-optical (T-CAM configuration in Section 4, directly supporting, for the first time, a wildcard bit operation of a logical "X" value in the optical domain [51]. By introducing Wavelength encoding in the search word, a 4-bit Matchline (ML) architecture is developed, capable of providing a unique identifying signal upon a match of the destination address with the stored prefix-entry [51]. Finally, we present a discussion on the future challenges that need to be addressed for migrating towards optical AL table architectures, bearing promises to directly resolve the AL in the optical domain that can significantly speed-up AL-speeds in high-end router architectures.

2. Electronic Address Look-Up Memory Architectures

Internet routers forward the data packets of an incoming port to an outgoing port based on an AL comparison function of the destination address of the header. To achieve this, they are equipped with a hardware look-up table that maintains the RIB-list of the destination addresses and their associated outgoing ports for next-hop routing, as depicted in the AL memory table architecture of Figure 1a. The architecture comprises a two dimensional CAM table inter-connected to a two dimensional RAM table. The CAM table stores the prefix-list with the destination addresses, while the RAM table maintains the outgoing port. Each entry of the RIB prefix-list is stored in a mutli-bit memory line, widely known as CAM ML. Upon the arrival of a packet, the header containing the destination address bits is broadcasted to all MLs of the CAM table, where it gets bitwise compared with the contents of the CAM cells. When the stored word of a specific ML matches the incoming search bits of the destination address, a match-signal is generated at the output of the ML, activating the specific line; otherwise, the CAM ML is not enabled. The activated CAM ML can then be mapped though the intermediate encoder-decoder network to the associated RAM table line, where the next-hop routing information is stored, so as to retrieve the outgoing port of the data packet. An example operation of the AL table is described by the RIB table shown at the right side of Figure 1a, assuming an incoming packet designated with a destination address [0111] and an RIB-list with four entries, namely [001X],

[010X], [011X], and [10XX]. By broadcasting and bit-wise comparing the destination address with all the RIB-entries, the third prefix of [011X] matches the packet's destination address. This matching will in turn activate the third line of the RAM forwarding table where port C is stored, indicating the next hop to which the incoming packet has to be forwarded.

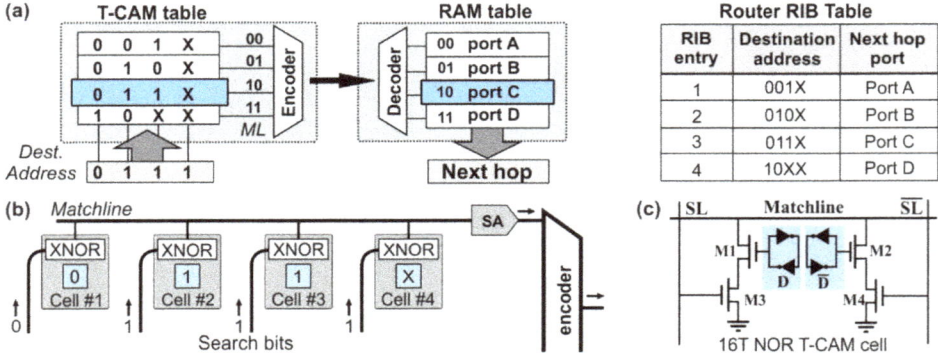

Figure 1. (**a**) Electronic AL memory architecture comprising a T-CAM table interconnected to a RAM table and the respective router Routing Information Base (RIB) list; (**b**) Logical circuit of an electronic Matchline architecture; and (**c**) the standard electronic 16T Not OR (NOR) T-CAM cell architecture.

A more detailed view of the block diagram describing the logical circuit operation of the 1×4 ML is shown in Figure 1b, comprising four parallel CAM cells. The CAM cells have their outputs combined at an inline Sense Amplifier (SA) and are then inter-connected to the encoder/decoder network for communication with the RAM table. Every CAM cell stores a bit of a RIB prefix entry and bears a logical XNOR gate for comparison with the incoming search bits of the destination address. If the search bit and the stored bit are equal, the logic XNOR gate transmits a signal of logic "1" value to the end of the respective ML. Equivalently, when all CAM cells of the line provide a logical match, the SA identifies the exact word match with the stored content and emits a proper signal towards the encoder. On the contrary, when there is a mismatch at any of the bits of the ML, the XNOR gate of the CAM cell provides a logical "0" value at the output, and de-activates the SA and the ML, denoting that a mismatch between the incoming destination address and the stored prefix has occurred. At this point, it is worth noting that MLs consist of Ternary T-CAM cells, in order to support ternary features with masked wildcard bits of "logical *X*", as used in subnet masks, rather than simpler Binary CAM cells that store only bits of logical "1" or "0".

The conventional architecture of the most typical 16T NOR-based electronic T-CAM cell is illustrated in Figure 1c [10], comprising two D-type flip-flops operating as storage cells and marked with a blue highlight. The first memory cell is responsible for storing the actual data-bit [0 or 1] and the second one stores the ternary state information for the "Care" or "Don't Care" state of the logical "*X*" value. Each memory cell is typically built on the configuration of two cross-coupled inverters, similar to 6T RAM cells [11], while another two pairs of transistors, (M1, M3) and (M2, M4), form short–circuits to the ground. This configuration ensures that at least one pulldown path from the ML down to the ground exists at any time during a mismatch between the search word and the stored data. On the contrary, a match between the search word bit and the stored data disables both pulldown paths, disconnecting the ML from the ground and feeding its output to the SA. If this matching happens for all the cells of the ML, a NOR logic operation between all CAM outputs, performed at the SA, identifies the exact match of the packet's destination address with the certain RIB prefix entry.

3. Monolithic Photonic Integration for Optical Memories

To this end, photonic integration processes have matured to the point where a wealth of optical FF and RAM memory configurations have been successfully demonstrated [33–41], with most of these demonstrations relying on the mature SOA switching technology, owing to its high-gain, high-speed, and high-yield performance characteristics [2]. In this paper, our analysis draws from the architecture of an FF memory with two cross-coupled asymmetric SOA-MZI switches, as illustrated in Figure 2a, which was initially developed in [37] and more recently theoretically investigated in the time and frequency domain in [43]. Each SOA-MZI switch features one SOA (SOA1 and SOA2) at the lower branch and a phase shifting element ($\Phi1$ and $\Phi2$) at the upper branch for controlling the biasing conditions of the interferometer. The two SOA-MZIs (SOA-MZI1 and SOA-MZI2) are each powered by a weak Continuous Wavelength (CW) input signal at wavelengths $\lambda1$ and $\lambda2$, respectively, while their Unswitched output ports (U-ports) are interconnected through a common coupling waveguide path. This symmetric configuration of the two SOA-MZI switches allows for a master-slave operation with the U-output of the master switch controlling the operating condition of the opposite slave switch and blocking its transmission at the U-port. Owing to the symmetric configuration, the roles of the master and slave switch are interchangeable, allowing the FF state to be defined by the wavelength of the emitted signal outputs at either one of the Switched ports (S-ports). The FF state can be monitored through the respective FFOut1 and FFOut2 output ports in terms of high power levels of $\lambda1$ and $\lambda2$ optical signals, respectively.

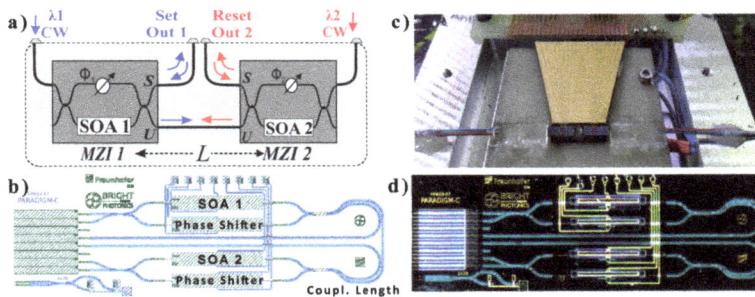

Figure 2. (**a**) All-optical Flip-Flop architecture and operation; (**b**) Mask-layout of the fabricated monolithic InP Flip-Flop; (**c**) Image of the packaged Flip-Flop device with electrical/optical connections and TEC element; and (**d**) Microscope image of the fabricated Flip-Flop device.

The state of the FF can be optically controlled by external *Set/Reset* (SR) pulses fed through the S-ports of the SOA-MZIs. When a high *Set* or *Reset* pulse is injected at the master SOA-MZI, it will block its light-transmission and set it in the slave condition regardless of the previous state of the FF, allowing for the opposite SOA-MZI to recover and become the master. In this arrangement, the two logical values of the data-bit stored in the FF-memory, logical "1" and "0", can be associated with the high optical power of the wavelength-signal emerging at the two outputs. The present FF memory configuration was initially demonstrated as a hybridly integrated module in [37] with the use of silica-on-silicon integration technology, exhibiting a total footprint of 45×12 mm^2 and coupling length between the two SOAs of 2.5 cm, which was theoretically shown to be the main speed determining factor of the memory architecture [42,43].

Following the conclusions drawn by the underlying theory on the critical performance parameters [42,43], an integrated version of the SR FF was presented using library-based components of a generic monolithic InP platform [33], to benefit from an integration technique that offers the possibility to fabricate multiple active and passive photonic components on a single chip at a close proximity. The two SOA-MZI switches were fitted in a die footprint-area of 6×2 mm^2 and cross-coupled together through a 5 mm-long coupling waveguide. In this configuration, the intra FF coupling waveguide was 5 mm-long, reducing the total footprint by two orders of magnitude, compared to the previous hybridly-integrated FF implementation [37]. The mask file of the Photonic Integrated Chip (PIC) is illustrated in Figure 2b. The input MMI couplers of the asymmetric MZIs featured a cross/through coupling ratio of 70/30, with respect to the SOA elements, and the MMIs between the two SOA-MZIs featured a coupling ratio of 50/50. The two SOAs featured an active length of 1 mm, while the biasing of the SOA-MZIs was achieved through the respective current injection phase shifters. Electrical routing of the metal wires connected the anodes and cathodes of the SOAs and the phase shifters to 100×100 μm^2 pads at the upper edge of the chip. The chip featured an array of eight zero-angled Spot Size Converters (SSC) with a pitch of 127 μm for optical I/O connectivity.

The monolithic InP FF-chip was fabricated by Fraunhofer Heinrich-Hertz-Institut (HHI) within a Multi-Project Wafer (MPW) run of the PARADIGM project funded by the European Commission and was later fully packaged in terms of the optical and electrical contacts for system level characterization. The chip was mounted on top of a ceramic sub-mount module, equipped with a thermistor and a Peltier element for temperature stability, while an 8-I/O fiber array was permanently glued to the I/O left facet of the chip. The metal pads of the chip were wire-bonded to the gold-plated ceramic mount, through gold wires, which also facilitated further connectivity with a Printed Circuit Board (PCB), where a 26-pin D-connector was mounted. The fully-packaged chip can be seen in Figure 2c and a microscope image of the PIC is shown in Figure 2d. The FF device was fully characterized in terms of its active components, i.e., the SOAs and the phase shifters, and was used to experimentally demonstrate a successful *Set-Reset* FF operation at 10 Gb/s [33], highlighting the potential to store and write data directly in the optical domain.

The present monolithic FF architecture served as the memory element of a more complex optical RAM cell configuration, capable delivering Read/Write and Block Access operations. To achieve this, the optical FF is combined with an SOA-MZI optical Access Gate (AG) in a cascade configuration, as shown in Figure 3a. The AG is responsible for granting access to the RAM cell by allowing the data to be either written to or read from the FF each time. The input Data signals are connected to the input of the SOA-MZI AG, and subsequently to an Arrayed Waveguide Grating (AWG) demultiplexer that drives each wavelength to the input/output ports of the FF-memory. The control signal is fed to the upper branch of the SOA-MZI AG, to induce Cross-Phase Modulation (XPM) phenomena. This allows the *bit* and $\overline{bit}$ signals to pass through the AG switch and emerge at the U-port, when there is no control pulse present, or at the S-port when a control pulse is present. RAM cell operations are then defined based on the values of the logic pulses of the *Inverted Access* signal and the *bit* and $\overline{bit}$ signals, which are fed to the RAM I/O Data port and Access port.

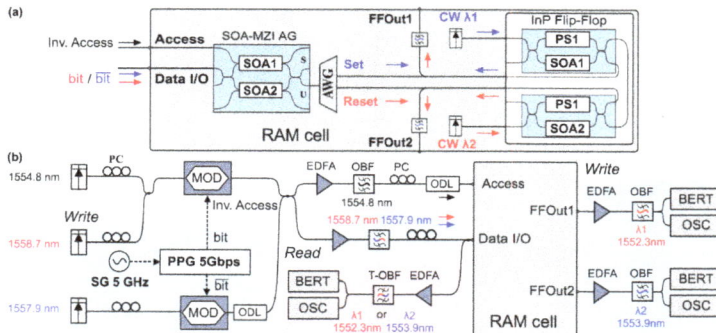

Figure 3. (**a**) Optical RAM cell architecture; and (**b**) Experimental layout used for an evaluation of the RAM Read/Write operation.

During the Write operation, when access is granted to the memory, the complementary *bit* and $\overline{bit}$ signals carry the incoming data word wavelength encoded on two different wavelengths, while the *Inverted Access* signal features a logical "0" pulse. Then, the external *bit* and $\overline{bit}$ signals pass through the AG U-port, and propagate towards the right side, where they are demultiplexed through the AWG and fed to the two inputs of the FF, acting as *Set* and *Reset* signals. During the Read operation, the *Inverted Access* signal again features a logical "0" pulse, while no external data are transmitted to the RAM cell. In this case, the complementary FF output signals, propagating from the FF towards the AG on the left side, are multiplexed in the AWG and fed to the AG through the U-port, so as to emerge at the output of the RAM cell. When access to the memory is blocked, the *Inverted Access* signal features a logical "1" pulse, which enters the SOA-MZI AG as the control signal, switching the complementary data signals to the S-port and blocking communication with the outer world, allowing for the FF-memory to retain its logic state and memory content.

The experimental setup used to evaluate the complete RAM functionalities is illustrated in Figure 3b. A signal generator (SG) was used to drive a Programmable Pattern generator (PPG) at 5 GHz. The PPG drives two Ti: LiNbO$_3$ modulators using complementary bit-patterns in order to produce 5 Gb/s $2^7 - 1$ Pseudorandom Binary Sequences (PRBS) Non-Return-to-Zero (NRZ) signals. One modulator is responsible for producing the *Inverted Access* signal at 1554.8 nm and the *bit* signal at 1558.7 nm, and the other is used for the $\overline{bit}$ signal at 1557.9 nm. For the evaluation of the Write operation, the three signals were coupled together to form the complementary *bit* and $\overline{bit}$ pair and one access signal, while the two coupler outputs were amplified using erbium-doped fiber amplifiers (EDFA). The first branch incorporated a 0.6 nm 3 dB-bandwidth Optical Bandpass Filter (OBF) centered at 1554.8 nm to properly filter the *Inverted Access* signal, while the second branch incorporated a 1 nm 3 dB-bandwidth OBF centered at 1558.3 nm, to filter the *bit* and $\overline{bit}$ pair. The AWG used has a 0.65 nm 3-dB channel bandwidth. The stored logical value of the FF and the *Set/Reset* signals could be monitored through the auxiliary ports FFOut1 and FFOut2 at any time of the experiment, which were amplified in respective EDFAs and filtered by suitable Optical Bandpass Filters (OBPFs), before being analyzed by a digital Optical Sampling Oscilloscope (OSC) and Bit Error Rate Tester (BERT). The blue-highlighted areas of the setup were only used during the evaluation of the Write operation, while the red-highlighted part of the setup was only used during the evaluation of the Read operation. To evaluate the READ operation, the FF was set to one of its logic memory states each time by properly adjusting the external CW signal power-levels, meaning that either only $\lambda 1$ or $\lambda 2$ was the dominant wavelength of the FF, providing a high FF output power level at either the FFOut1 or FFOut2 port, respectively. A monitor branch was connected at the Data I/O port of the RAM cell comprising an EDFA as a preamplifier and a 0.6 nm 3 dB-bandwidth Tunable OBPF (T-OBPF) that can be tuned at either one or two in order to evaluate one of the two wavelengths. Polarization controllers (PC) were used at several stages of the setup to control signal polarization. Variable optical attenuators (VOA) were also used to properly adjust the power levels of the optical signals, while optical delay lines (ODLs) were employed to ensure signal decorrelation and bit-level synchronization among the signals.

The experimental results obtained from the 5 Gb/s operation can be seen in Figure 4. Figure 4a–g depict synchronized time traces/eye diagrams of the Write operation. Figure 4a shows the *Inverted Access* signal and Figure 4b,c show the *bit* and $\overline{bit}$ signals that were launched in the RAM cell, respectively. Figure 4d,e illustrate the *Set/Reset* signals originating from the incoming *bit* and $\overline{bit}$ signals, after the access-controlling operation of the AG, where it is clear that they only imprint the logical "1" *bit* and $\overline{bit}$ pulses, respectively, when there is no logical "1" pulse at the *Inverted Access* signal. The proof-of-principle of the Write operation is then verified by monitoring the FF stored content through the signals emerging at the respective FFOut1 and FFOut2 ports in Figure 4f,g, where it can be seen that the FF changes its logic state when there is an incoming *Set* or *Reset* pulse and maintains its state until the next *Set/Reset* pulse arrival. The eye diagrams of the FF output signals feature a recovery time of 150 ps and an Extinction Ratio (ER) of 6 dB.

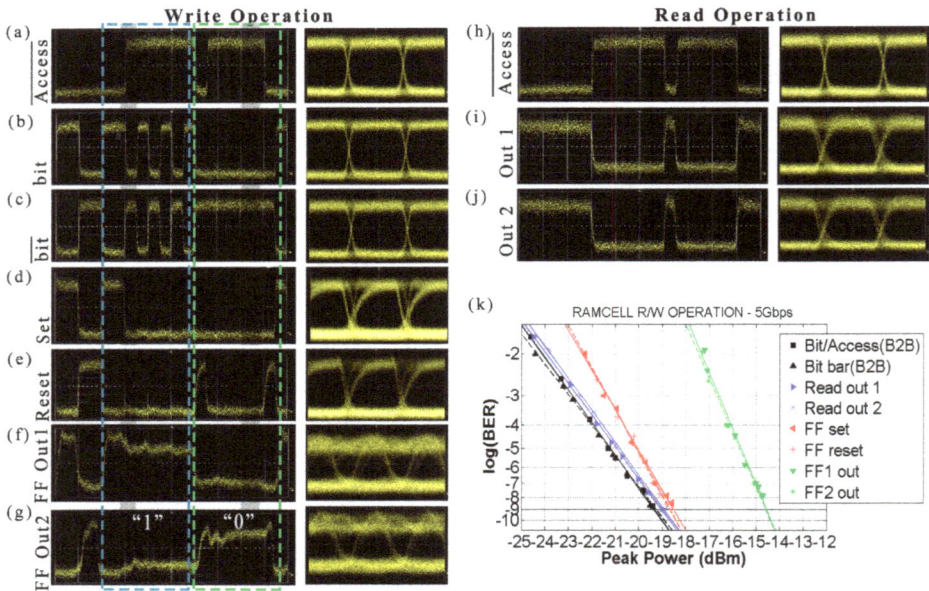

Figure 4. Experimental results of RAM cell demonstration: (**a–g**) Time traces (400 ps/div) and eye diagrams (50 ps/div) for the WRITE operation; (**h–j**) Time traces (400 ps/div) and eye diagrams (50 ps/div) for the READ operation; (**k**) BER measurements for the READ/WRITE operation at 5 Gb/s.

Figure 4h–j depict synchronized time traces and the respective eye diagrams for the Read operation. For this evaluation, the FF was set to one logic state and the data were then read by transmitting the *Inverted Access signal*, shown in Figure 4h. Figure 4i,j illustrate the time traces of the RAM cell output signals, after the random access controlling operation of the AG for both of the FF logic states. A high output value is only obtained when there is no *Inverted Access* signal pulse present, indicating that access is granted for the Read operation, while when there is a high optical power for the *Inverted Access* signal, the RAM output of the RAM cell features a logical "0" and access to the memory is thus blocked. The eye diagram of the read output features an ER of 9 dB.

Figure 4k shows the BER measurements obtained for both the Write and Read RAM cell functionalities, where, in all cases, error free operations are demonstrated. For the Write operation, the BER diagrams reveal a power penalty of 0.6 dB for the *Set* and *Reset* signals compared to the Back-to-Back measurements of the *bit* and $\overline{bit}$ signals, and in turn, 4.6 dB for the FFOut1 and FFOut2 signals of the data written to the RAM cell, while for the case of the Read operation, a power penalty of only 0.5 dB was obtained at the 10^{-9} condition. The results were obtained with a current injection of around 250 mA for each SOA, while when relying on four SOAs operating at 5 Gb/s, the resulting energy efficiency per bit is 400 pJ/bit.

4. Optical CAM Technology

In this section, we describe the recent developments towards developing optical CAM architectures, spanning from the first optical Binary CAM cell to a T-CAM cell layout and the design of multi-bit CAM ML architectures.

4.1. Experimental Demonstration of an Optical CAM Cell

The architecture of the optical Binary CAM cell is schematically illustrated in Figure 5a, comprising the optical FF memory connected to an SOA-MZI XOR logic gate. The optical FF stores the data of the CAM cell, while the XOR logic gate is used for comparing the incoming search bit of the destination address with the stored bit of the prefix. As described previously, the FF is powered by two wavelengths, $\lambda 1$ and $\lambda 2$, which are emitted at its two outputs, respectively. In order to achieve a content Comparison operation, one of the two wavelengths, carrying the stored bit information, here $\lambda 1$, is connected as the control signal to the SOA-MZI XOR gate, while the second control branch of the SOA-MZI is fed with the incoming search bit. The comparison result between the search bit and the FF-memory content is imprinted on a new wavelength, $\lambda 3$, that is fed as a CW probe signal at the SOA-MZI gate and emerges at its S-output port, forming the final CAM cell output signal. When the two signals feature equal bit-pulses and a match occurs between them, the $\lambda 3$-CW emerges at the U port of the SOA-MZI and a bit pulse of logical "0" is thus obtained at the S-port and the CAM cell output. On the contrary, when the search bit does not match the FF memory content, a differential π-phase shift is obtained at the SOA-MZI, and the $\lambda 3$-CW signal emerges at the S-port and a pulse of logical "1" is obtained at the CAM cell output. Regarding the update of the CAM memory content, a Write operation has to be performed for the FF by launching *Set*/*Reset* pulses through the respective CAM cell optical ports, so as to change its logic state.

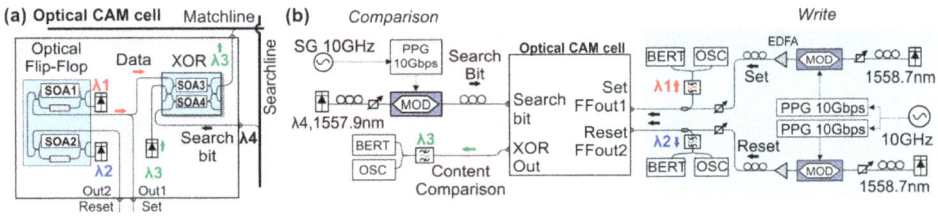

Figure 5. (**a**) Proposed architecture of an all-optical CAM cell; (**b**) Experimental setup for the demonstration of the Comparison operation and Write operation.

The proposed optical CAM cell architecture was experimentally investigated for both operations, i.e., content Comparison and Write operation at 10 Gb/s, using the experimental setup shown in Figure 5b. For the Comparison operation, the red-highlighted area of the setup was used. A 10 GHz SG was used to drive a PPG that modulated a LiNbO$_3$ modulator, in order to produce the 10 Gb/s NRZ $2^7 - 1$ PRBS signal at a $\lambda 4 = 1557.9$ nm wavelength, forming the search signal that is transmitted to the SOA-MZI XOR gate as the control signal and thus emulating the external search-bit. The comparison result is imprinted at the $\lambda 3 = 1548$ nm wavelength and the result is obtained at the respective SOA-MZI XOR output, after being filtered at a 0.6 nm 3 dB-bandwidth OBPF. For the Write operation, the blue-highlighted area of the setup was used. The 10 GHz SG was connected to a PPG, which then provided two 10 Gb/s NRZ data streams that modulated two LiNbO$_3$ modulators to generate the *Set*/*Reset* data steams at a 1558.7 nm wavelength. The logic state of the FF was recorded through the respective FFOut1/2 ports at $\lambda 1$ and $\lambda 2$, respectively, filtered by 0.6 nm 3 dB-bandwidth OBPFs centered at $\lambda 1$ and $\lambda 2$, which were then analyzed by an OSC and a BERT. EDFAs and VOAs were incorporated in both experimental setups for power loss compensation and power level management of the signals, while PCs were used to adjust the signal polarization states.

The experimental results obtained in the optical CAM cell demonstration can be seen in Figure 6. Figure 6a–d show the time traces and the respective eye diagrams obtained during the Content Comparison operation. The FF was manually set each time to one of the two states and studied for both stored FF data-bit of logical "0" and logical "1", shown in Figure 6a. For the first logic state, when the FF holds a logic value of "0", CW1 emerges at FFOut1 with a low power level and enters the

SOA-MZI XOR gate, while Figure 6b shows the time trace of the search bit signal acting as the second control signal of the XOR operation. Figure 6c shows the result of the XOR comparison, where it can be seen that a logic "1" pulse is only obtained when a "logic 1" pulse is present at the search bit, featuring the same data pattern and confirming an XOR operation with the FF state of logic "0". Equivalently, when the FF holds a logic value of "1", as shown in the second column of Figure 6a, and is compared with the input search bit trace, the obtained CAM cell output features an inverse logic bit pattern compared to the input search bit, as shown in Figure 6c, verifying the XOR proof-of-principle. The eye diagrams of the obtained XOR output signals of the CAM cell are shown in Figure 6d, featuring an average ER of 9.2 dB.

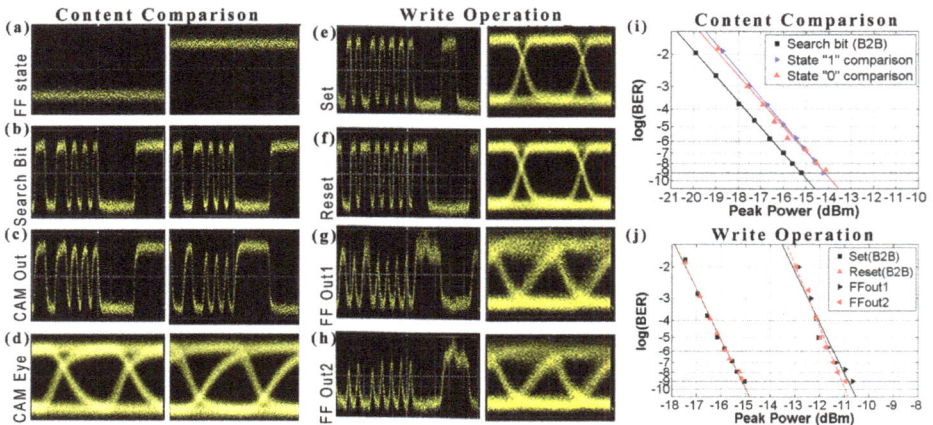

Figure 6. Experimental results of the optical CAM cell demonstration at 10 Gb/s: Comparison operation: time traces (200 ps/div) of (**a**) CAM cell stored bits; (**b**) search bits; (**c**) CAM outputs and (**d**) output eye diagrams (15 ps/div). Write Operation: Time traces at 10 Gb/s (300 ps/div) their corresponding eye diagrams (20 ps/div) of (**e**) the *Set*; (**f**) the *Reset*; (**g**) FFOut1 and (**h**) FFOut2; and BER measurements of (**i**) the Content Comparison and (**j**) the Write operation.

The results for the CAM cell Write mode functionality are presented in Figure 6e–h. Figure 6e,f show the *Set* and *Reset* signals and their corresponding eye diagrams, that were sent to the CAM cell in order to control the state of the FF. Figure 6g,h show the two stored bit values of the FF as a recorder at its two outputs, where it can be seen that the FF changes state upon the arrival of a Set (or Reset) pulse with a recovery time of 80 ps and the memory is then maintained until the arrival of the next opposite pulse. The two FF output signals clearly feature open eye diagrams, revealing an average ER of 6.5 dB. The two operations of the CAM cell were also evaluated with the aid of BER measurements and the results are shown in Figure 6i,j for the content comparison and the write functionality, respectively. An error free operation was achieved at the 10^{-9} operating condition for both states of the FF and for both CAM cell functionalities. The BER diagram in Figure 6i reveals a successful content comparison at 10 Gb/s with a power penalty of 1 dB, associated mainly with the signal degradation induced by the optical XOR gate, while the BER diagram of Figure 6j for the Write functionality exhibits a power penalty of 4 dB, owing to the dynamic operation of the FF. For the CAM cell experimental demonstration, the four SOAs were again operated at a current of 250 mA each, leading to an energy efficiency of 200 pJ/bit at a 10 Gb/s operational speed.

4.2. Ternary CAM Cell and Matchline Architecture

In this section, we move from the all-optical B-CAM cell to the presentation of the all-optical T-CAM cell architecture and its interconnection in an all-optical T-CAM row arrangement targeted for use in AL tables such as the one presented in Figure 1a, enabling the essential subnet-masked operation needed in modern router applications. The proposed optical T-CAM cell architecture comprises two optical FFs and an optical XOR gate; the 1st FF is used for storing the actual T-CAM cell contents and the 2nd FF for implementing the "X" state support. The XOR gate, on the other hand, enables the T-CAM cell search capability. Moving to the complete row arrangement, the proposed all-optical T-CAM row architecture comprises an indicative number of four T-CAM cells followed by a novel WDM-encoded ML design, providing a comparison operation for complete 4-bit optical words. In order to achieve this, an AWG is utilized to multiplex all the T-CAM cell XOR outputs in a common multi-wavelength row output signal that determines whether a success comparison result is achieved throughout the complete T-CAM row.

Figure 7 presents the proposed all-optical T-CAM cell consisting of two FF modules and one XOR gate. The XOR gate is necessary for realizing the comparison operation between the *search-bit* and the value stored in the T-CAM cell. The lower FF is named XFF and is necessary for implementing the third state "X", while the upper FF is named T-CAM Content FF (TCFF) and stores the actual content that can be either a logical "0" or "1". When a subnet-masked operation is desired, the XFF's content equals 0, implying that the TCFF respective content has to be ignored. As such, the respective XOR operation does not take into account the TCFF content and the comparison result is equal to a logical "0", independently of the value of the *search-bit*. On the contrary, in the case where the TCFF value has to be taken into account, the XFF content is equal to a logical "1" and the XOR output depends upon the comparison between the TCFF value and the respective *search-bit*.

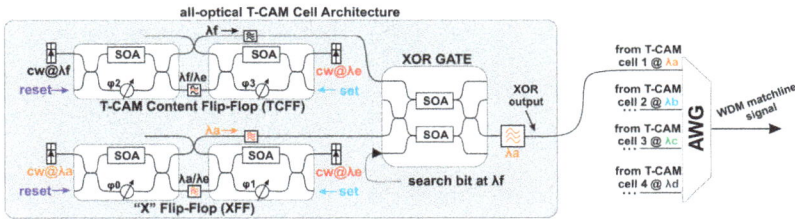

Figure 7. All-optical T-CAM cell architecture with two FFs (TCFF & XFF) and a XOR gate and a T-CAM row's AWG multiplexer for four indicative T-CAM cells.

For both FFs, the previously described *Set/Reset* pulse mechanism is used in order to switch between the two possible logical states. The XFF and TCFF are each powered by two CW laser beams: λe is used as the input signal at the right-side switches of both XFF and TCFF, while a CW signal at λa and λf is launched as the input signal at the left-side SOA-MZI switches of the XFF and TCFF, respectively. As such, the content of the XFF and the TCFF gets encoded on λa and λf wavelengths as the FF output signals, respectively. The XFF output signal is then fed as the input signal at the XOR gate, after being filtered in an OBF. On the other side, the TCFF output at λf enters the XOR gate as the control signal of the upper-branch SOA. The lower branch SOA of the XOR gate is being fed with the input *search bit* that acts as the second control signal. In this way, the TCFF output and *search bit* values get logically XORed and the comparison result gets imprinted on the XFF output signal at λa, which is used as the XOR input. Whenever the T-CAM cell is in the "X" state, the XFF output is equal to a logical "0", resulting in a logical "0" at the final XOR output, irrespective of the TCFF and *search bit* values. On the contrary, the final XOR output depends on the comparison result between the TCFF output and *search bit* values when the XFF output equals a logical "1": when both the TCFF output and *search bit* signals have the same value, the XOR output is "0", while in the opposite case, i.e., when they have different values, the XOR output equals "1" and is imprinted on the XFF output at λa.

By assigning a different wavelength for carrying the optical XOR output at every individual T-CAM cell within a row, all four of the T-CAM cell outputs can be combined at the row output by using an AWG multiplexer, as presented at the right side of Figure 7; λa through λd are used for the different cell outputs, while λe, λf, and the wavelengths used for the *Set/Reset* signals are employed in all T-CAM cells. This leads to a WDM-encoding scheme that produces the corresponding *ML signal* at the final row output. In this way, an *ML signal* of a logical value "*0*" indicates a completely matched comparison result since all the individual XOR outputs will be equal to a logical "*0*". On the contrary, a non-zero optical power level obtained at the encoder input indicates that at least one individual XOR output produces a comparison miss, denoting a non-completely matched row. A more generic representation of the proposed T-CAM row architecture is presented in Figure 8.

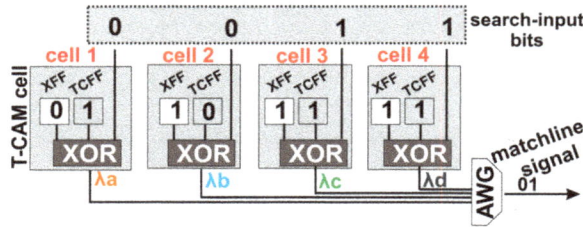

Figure 8. T-CAM row architecture comprising an indicative number of four T-CAM cells.

Figure 9 presents the simulation results of the T-CAM row architecture for both Search and Write operations and at a line-rate of 10 Gb/s. The simulation models have been developed using the VPI Photonics suite and both the XOR gate and FF models are based on experimentally verified building blocks. The SOA model used in both the XOR gates and FFs is identical to the one presented and experimentally validated in [52]. The wavelengths used in the four-cell arrangement of Figure 7 are equal to: λa: 1564.19 nm, λb: 1562.56 nm, λc: 1559.31 nm, λd: 1557.36 nm, λe: 1554.78 nm, λf: 1546.12 nm, *Set*: 1548.35 nm, and *Reset*: 1551.88 nm.

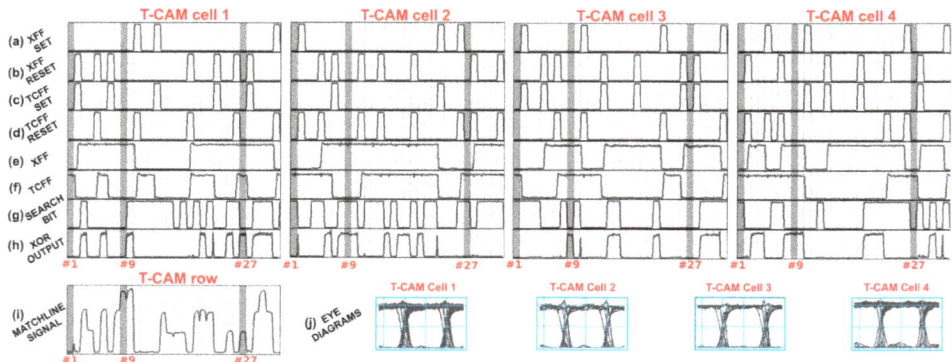

Figure 9. 10 Gbps simulation results for the T-CAM row architecture of Figure 8: (**a**) the XFF Set; (**b**) the XFF Reset; (**c**) the TCFF Set; (**d**) the TCFF Reset; (**e**) the XFF; (**f**) the TCFF; (**g**) the Search Bit; (**h**) the XOR output; (**i**) the Matchline output signals, all with a time scale of 100 ps/div for traces and (**j**) the eye diagrams of the four TCAM cell outputs at 50 ps/div.

Figure 9a,b illustrate the *Set/Reset* pulse traces that are fed into the XFFs and determine whether the XFF has to define an "X" state for the T-CAM cell. Figure 9c,d illustrate the *Set/Reset* pulse traces that are fed into the TCFFs of the four cells dictating the logical content of every TCFF. Figure 9e depicts the XFF output signal, while Figure 9f illustrates the TCFF content transitions for every T-CAM cell. As can be seen in both Figure 9e,f, a successful bit storage operation is achieved according to the respective *Set*/*Reset* pulse traces; the presence of a *Set* pulse leads to an FF content transition to the 0 logical state, while the presence of a Reset pulse leads to an FF content transition to the logical state of 1. Figure 9g presents the *search-bit* pulse traces that are fed into the XOR gates of the four T-CAM cells as parallel streams in order to be compared with the respective T-CAM cell contents. The *search-bit* pulse traces are NRZ $2^7 - 1$ PRBS at a line-rate of 10 Gb/s. Figure 9h shows the XOR output signals that also form the T-CAM cell outputs and Figure 9i illustrates the power level of the final ML signal that is produced at the row output and just after the AWG multiplexer. As can be noticed, this is a multilevel signal, with every different power level corresponding to a different number of bit-level search misses. When all T-CAM cells match the four bits of the incoming *search-input* signal, no optical power is recorded at the AWG output.

A successful ML operation of the complete T-CAM row can be verified for the entire pulse traces used as the four parallel *search bit* sequences. Three representative examples at the timeslots #1, #9, and #27 have been highlighted in order to facilitate the understanding of the T-CAM row performance in different situations. In the example of timeslot #1, all four T-CAM cells are in the "X" state since all respective XFF outputs are equal to "0", which finally results in an XOR output of "0", regardless of the TCFF and *search-bit* values. As expected, the final ML signal at timeslot #1 is also equal to "0", corresponding to a complete match between the T-CAM row and the *search-input* contents. Within timeslot #9, none of the T-CAM cells are in the "X" state since all XFF outputs are equal to a logical 1. For T-CAM cell #1, the XOR output is equal to a logical "0" since the TCFF and the *search-bit* content are equal. However, for the remaining three T-CAM cells, the respective XOR outputs are equal to a logical "1", denoting the different content between the corresponding TCFF and *search-bit* signals. The presence of three optical pulses at different wavelengths but within the same timeslot #9 designates that the optical power obtained at the AWG output will equal the sum of the power levels of the three individual pulses, obviously leading to an ML signal with non-zero power that indicates a non-perfectly matched search operation. In the example of timeslot #27, T-CAM cells #2 and #4 are in the "X" state since their XFF content equals a logical "0". As such, the respective XOR outputs are also equal to "0" and this happens even in the case of T-CAM cell #4, where the TCFF content and its respective *search-bit* are different. Regarding cell #3, the XOR output equals "0" because both the TCFF content and the *search-bit* are equal. However, cell #1 has an XOR output of "1" since TCFF content and the respective *search-bit* have different values. This single optical pulse obtained as the result of the comparison along the entire T-CAM row is then also translated into a non-zero power level at the AWG output, again suggesting a non-matched row, while Figure 9j presents clearly open eyes for all four T-CAM cells with an ER of 13 dB. The energy efficiency of the TCAM cell is 300 pJ/bit.

5. Future Challenges and Discussion

The presented multi-wavelength optical subsystems provide the necessary constituents for synthesizing a new design roadmap for a photonic AL memory architecture that can reap the unique benefits of a high-bandwidth and low-power consumption offered by optical technology. The use of discrete SOA-MZIs interconnected to fiber-pigtailed monolithic integrated FF-devices allows us to overcome the critical speed determining factor of the large intra-FF coupling distance [42,43] and facilitates the full characterization of each RAM cell and CAM cell independently, without any speed limitation, when unidirectional data-communication is employed. However, the latency of the fiber-network of the complete optical ML architecture introduces some latency in the overall destination address resolution, when the content comparison output of the CAM cell has to be propagated to the RAM cell for a Read operation. By reaping the benefits of mature photonic integration multiple photonic components per chip [2], a fully integrated Matchline could enable studying all RAM and CAM functionalities simultaneously, with a shorter latency for the destination address resolution operation, which could be further improved by incorporating gain-accelerating techniques [53]. Meanwhile, for a simultaneous synchronous write operation at the RAM cell and CAM cell of the envisioned optical ML, the write operation should be performed at 5 Gb/s due to the cascaded switching operation at the AG, which is already an order of magnitude faster than the respective electronic memory speed [6,54], while faster memory updates and network changes are subject to the use of higher speed SOAs. Drawing from our presented initial experimental timing tests with short rise/fall times of 90 ps, optical CAM memories can also significantly reduce the reconfiguration times towards enabling rapid updates of the RIB-memory.

Our proposed scheme has relied on power-hungry 1 mm-long SOAs of a generic foundry that necessitate external currents of 250 mA and a power consumption of 0.5 W for each SOA for proof-of-concept demonstration purposes, resulting in an optical multi-bit multi-wavelength ML architecture with a power efficiency of 300 pJ/bit per clock cycle when operating at 10 Gb/s. This value can be reduced by orders of magnitudes when shifting to more sophisticated low-power III/V-on-SOI photonic crystal technologies with nm-scale dimensions and power consumptions of a few mW [55], towards energy efficiencies of a few fJ/bit, comparable to electronics [6,20,21]. Meanwhile the use of the envisioned high-speed multi-bit optical ML architectures technology provides a possible path towards circumventing the use of power-hungry cost expensive power conversion at SERDES equipment, that can allocate up to half of the power consumption of a low power transceiver [24]. Additional power consumption benefits may also be obtained, when shifting to higher bitrates beyond 10 Gb/s [43] or even when exploiting the wavelength dimension for a single Access Gates shared among the multiplexed outputs of multiple optical RAMs [45].

The use of optical AL memories can also benefit from the inherent support of WDM techniques for enhancing wavelength parallelism and reaping the advantages of a new multiplexing capability that is so far not feasible in electronic AL memories. Towards a more practical application of a fully integrated optical ML architecture suitable for handling IPv4 addresses, and considering the typical table sizes of electronic memories with 4K entries [56], a row-capacity of 32-bit would be required, while for compatibility with next generation IPv6 addresses, a 128-bit memory would be required. In this case, the ML architecture will utilize 32 or 128 wavelengths, respectively, one per TCAM cell output, similar to the wavelength addressable large-scale photonic integrated photonic memories of 128-bits [30] or the 128-wavelengths of the optically connected RAM architectures for High Performance Computers [57], as well as an AWG with a similarly high channel count [1]. The assignment of different wavelengths at the various T-CAM cells of an ML certainly adds an extra degree of freedom to system-designs for more efficient architectures, especially when considering multiple simultaneous WDM memory schemes with parallel wavelengths written to the T-CAM cells of an ML for fast memory updates. This can be of utter importance with the emergence of SDN and OpenFlow architectures that enforce a dynamic network operation with frequent updates of network topologies and multiple real time changes in the RIB-list [25,26]. Although the SDN controllers are increasingly optimized for swift policy updates, the T-CAM tables remain yet unoptimized for fast updates [27,28], which may trigger hundreds to thousands of table entry-moves and write-memory operations [26]. Measurements on the timings of such AL-table updates have revealed a few hundreds of ms-long response times [25–28], as AL-updates need to be organized in electronic T-CAM tables with sequential time-multiplexed memory accesses through a memory-bus of a limited bandwidth. In this case, the optical nature of the proposed T-CAM may facilitate wavelength multiplexed memory access schemes in order to perform multiple simultaneous Write operations at 10 Gb/s, either when updating the prefix-list of the optical CAM-table or the outgoing ports stored in the optical RAM table, providing manifold improvements in the AL memory throughput.

When developing programmable low power optical AL-memory architectures, further system benefits could potentially be obtained by the introduction of the programmability of the optical memory content [31] or the use of a non-volatile photonic integrated memory [32]. However, apart from the optimization of the memory cell, additional developments are still required, as, e.g., the development of a proper input WDM interconnect network, capable delivering the input bit in vertical CAM-column arrangements, known as Searchlines, or the intermediate encoder/decoder interconnect network and peripheral sub-circuits that undertake the communication between the CAM-table and the RAM table, potentially taking into account the case of priority encoding, when multiple prefix matches are formed by the use of ternary bits. These next steps could potentially follow similar concepts as the ones suggested by the all-passive wavelength-based Column Address Selector [45], the peripheral circuitry [46], and/or the optical tag comparator [47] of the more detailed studies in the area of optical RAM architectures.

6. Conclusions

A novel WDM optical CAM bank architecture is presented, comprising a list of ML memory architectures capable of performing a content Comparison at 10 Gb/s for fast Address Look-Up (AL) operations. The presented architecture is built on the first optical Binary and Ternary CAM cell architectures, as alternatives to the ubiquitous optical Flip-Flops and optical Random Access Memory (RAM) cells with location-based memory addressing. The proposed ML architecture also reveals the potential for a multi-wavelength operation for the full exploitation of wavelength encoding, paving the way for multiple parallel WDM write access operations, suggesting manifold improvements in the programmability and reconfigurability of AL memories.

Acknowledgments: This work has been supported by the FP7-PEOPLE-2013-IAPP-COMANDER project (Contract No. 612257). The EU-FP7-ICT-PARADIGM project (Contract No. 257210) is acknowledged for the FF chip design and fabrication. The authors would also like to acknowledge Ronald Broeke and Francisco Soares for the chip fabrication and Tolga Tekin for the packaging.

Author Contributions: C.V. and S.P. designed and performed the experiments, P.M. designed and performed the simulations, A.M. and N.P. conceived the initial ideas. C.V., S.P. and P.M. wrote the paper and A.M. and N.P. reviewed the results and the manuscript.

Conflicts of Interest: The authors declare no conflict of interest.

References

1. The Internet of Things How the Next Evolution of the Internet Is Changing Everything. Available online: http://www.cisco.com/c/dam/en_us/about/ac79/docs/innov/IoT_IBSG_0411FINAL.pdf (accessed on 10 May 2017).
2. Smit, M.; Leitjens, X.; Ambrosius, H.; Bente, E.; Tol, J.; Smalbrugge, B.; Vries, T.; Geluk, E.J.; Bolk, J.; Veldhoven, R.; et al. An introduction to InP-based generic integration technology. *IOP Semicond. Sci. Technol.* **2014**, *29*, 1–41. [CrossRef]
3. Winzer, P. Scaling Optical Fiber Networks: Challenges and Solutions. *Opt. Photonics News* **2015**, *26*, 28–35. [CrossRef]
4. Ballani, H.; Francis, P.; Cao, T.; Wang, J. Making routers last longer with ViAggre. In Proceedings of the 6th USENIX Symposium on Networked Systems Design and Implementation, Boston, MA, USA, 22–24 April 2009.
5. Available Pool of Unallocated IPv4 Internet Addresses Now Completely Emptied. Available online: https://www.icann.org/en/system/files/press-materials/release-03feb11-en.pdf (accessed on 10 May 2017).
6. Nii, K.; Amano, T.; Watanabe, N.; Yamawaki, M.; Yoshinaga, K.; Wada, M.; Hayashi, I. A 28 nm 400 MHz 4-Parallel 1.6 Gsearchs 80 Mb Ternary CAM. In Proceedings of the IEEE International Solid-State Circuits Conference, San Francisco, CA, USA, 9–13 February 2014.
7. Growth of the BGP Table—1994 to Present. Available online: http://bgp.potaroo.net/ (accessed on 10 May 2017).
8. Arianfar, S.; Nikander, P.; Ott, J. On content-centric router design and implications. In Proceedings of the Re-Architecting the Internet Workshop, Philadelphia, PA, USA, 30 November 2010.
9. Ruiz-Sanchez, M.A.; Biersack, E.W.; Dabbus, W. Survey and Taxonomy of IP Address Lookup Algorithms. *IEEE Netw.* **2001**, *15*, 8–23. [CrossRef]
10. Pagiamtzis, K.; Sheikholeslami, A. Content-addressable memory (CAM) circuits and architectures: A tutorial and survey. *IEEE J. Solid State Circuits* **2006**, *41*, 712–727. [CrossRef]

11. Athe, P.; Dasgupta, S. A comparative study of 6T, 8T and 9T decanano SRAM cell. In Proceedings of the IEEE Symposium on Industrial Electronics & Applications, Kuala Lumpur, Malaysia, 4–6 October 2009.
12. Chisvin, L.; Duckworth, R.J. Content-Addressable and Associative Memory: Alternatives to the Ubiquitous RAM. *Computer* **2002**, *22*, 51–64. [CrossRef]
13. Shah, D.; Gupta, P. Fast Updating Algorithms for TCAMs. *IEEE Micro* **2001**, *21*, 36–47. [CrossRef]
14. Yang, B.D.; Kim, L.S. A Low-Power CAM Using Pulsed NAND–NOR Match-Line and Charge Recycling Search-Line Driver. *IEEE J. Solid State Circuits* **2005**, *40*, 1736–1744. [CrossRef]
15. Yang, B.D.; Lee, Y.K.; Sung, S.W.; Min, J.J.; Oh, J.M.; Kang, H.J. A Low Power Content Addressable Memory Using Low Swing Search Lines. *IEEE Trans. Circuits Syst. I* **2011**, *58*, 2849–2858. [CrossRef]
16. Kasai, G.; Takarabe, Y.; Furumi, K.; Yoneda, M. 200 MHz/200 MSPS 3.2 W at 1.5 V Vdd, 9.4 Mbits ternary CAM with new charge injection match detect circuits and bank selection scheme. In Proceedings of the IEEE Custom Integrated Circuits Conference, San Jose, CA, USA, 24–26 September 2003.
17. Hayashi, I.; Amano, T.; Watanabe, N.; Yano, Y.; Kuroda, Y.; Shirata, M.; Dosaka, K.; Nii, K.; Noda, H.; Kawai, H. A 250-MHz 18-Mb Full Ternary CAM with Low-Voltage Matchline Sensing Scheme in 65-nm CMOS. *IEEE J. Solid State Circuits* **2013**, *48*, 2671–2680. [CrossRef]
18. Moradi, M.; Qian, F.; Xu, Q.; Mao, Z.M.; Bethea, D.; Reiter, M.K. Caesar High Speed and Memory Efficient Forwarding Engine for Future Internet Architecture. In Proceedings of the ACM/IEEE Symposium on Architectures for Networking and Communications Systems, Oakland, CA, USA, 7–8 May 2015.
19. Jiang, W.; Wang, Q.; Prasanna, V.K. Beyond TCAMs: An SRAM-Based Parallel Multi-Pipeline Architecture for Terabit IP Lookup. In Proceedings of the IEEE INFOCOM Conference on Computer Communications, Phoenix, AZ, USA, 13–18 April 2008.
20. Jeloka, S.; Akesh, N.B.; Sylvester, D.; Blaauw, D. A 28 nm Configurable Memory TCAM BCAM SRAM Using Push Rule 6T Bit Cell Enabling Logic in Memory. *IEEE J. Solid State Circuits* **2016**, *51*, 1009–1021. [CrossRef]
21. Arsovski, I.; Hebig, T.; Dobson, D.; Wistort, R. A 32 nm 0.58-fJ/Bit/Search 1-GHz Ternary Content Addressable Memory Compiler Using Silicon-Aware Early-Predict Late-Correct Sensing With Embedded Deep-Trench Capacitor Noise Mitigation. *IEEE J. Solid State Circuits* **2013**, *48*, 932–939. [CrossRef]
22. Kilper, D.C.; Atkinson, G.; Korotky, S.K.; Goyal, S.; Vetter, P.; Suvakovic, D.; Blume, O. Power trends in communication networks. *IEEE J. Sel. Top. Quantum Electron.* **2011**, *17*, 275–284. [CrossRef]
23. Tucker, R.S.; Baliga, J.; Ayre, R.; Hinton, K.; Sorin, V.W. Energy Consumption in IP Networks. In Proceedings of the European Conference on Optical Communication, Brussels, Belgium, 21–25 September 2008.
24. Audzevich, Y.; Watts, P.; West, A.; Mujumdar, A.; Crowcroft, J.; Moore, A. Low power optical transceivers for switched interconnect networks. In Proceedings of the International Conference on Advanced Technologies for Communications, Ho Chi Minh City, Vietnam, 16–18 October 2013.
25. Hey, K.; Khalid, J.; Gember-Jacobson, A.; Das, S.; Akella, A.; Erran, L.L.; Thottan, M. Measuring Control Plane Latency in SDN-enabled Switches. In Proceedings of the 1st ACM SIGCOMM Symposium on Software Defined Networking Research, Santa Clara, CA, USA, 17–18 June 2015.
26. Karam, R.; Puri, R.; Ghosh, S.; Bhunia, S. Emerging Trends in Design and Applications of Memory-Based Computing and CAMs. *IEEE Proc.* **2015**, *103*, 1311–1330. [CrossRef]
27. Katta, N.; Alipourfard, O.; Rexford, J.; Walker, D. CacheFlow: Dependency-Aware Rule-Caching for Software-Defined Networks. In Proceedings of the Symposium on SDN Research, New York, NY, USA, 14–15 March 2016.
28. Wen, X.; Yang, B.; Chen, Y.; Errann, L.L.; Bu, K.; Zheng, P.; Yang, Y.; Hu, C. RuleTris: Minimizing Rule Update Latency for TCAM-based SDN Switches. In Proceedings of the International Conference on Distributed Computing Systems, Nara, Japan, 27–30 June 2016.
29. Koonen, A.M.J.; Yan, N.; Olmos, J.V.; Monroy, I.T.; Peuchert, C.; Breusegem, E.V.; Zouganeli, E. Label-Controlled Optical Packet Routing-Technologies and Applications. *IEEE J. Sel. Top. Quantum Electron.* **2007**, *13*, 1540–1550. [CrossRef]
30. Kuramochi, E.; Nozaki, K.; Shinya, A.; Takeda, K.; Sato, T.; Matsuo, S.; Taniyama, H.; Sumikura, H.; Notomi, M. Large-scale integration of wavelength-addressable all-optical memories on a photonic crystal chip. *Nat. Photonics* **2014**, *8*, 474–481. [CrossRef]
31. Song, J.F.; Luo, X.S.; Lim, A.E.J.; Li, C.; Fang, Q.; Liow, T.Y.; Jia, L.X.; Tu, X.G.; Huang, Y.; Zhou, H.F.; et al. Integrated photonics with programmable non-volatile memory. *Sci. Rep.* **2016**, *6*. [CrossRef] [PubMed]

32. Rios, C.; Stegmaier, M.; Hosseini, P.; Wang, D.; Scherer, T.; Wright, D.C.; Bhaskaran, H.; Pernice, W.H.P. Integrated all-photonic non-volatile multi-level memory. *Nat. Photonics* **2015**, *9*, 725–732. [CrossRef]
33. Pitris, S.; Vagionas, C.; Kanellos, G.T.; Kisacik, R.; Tekin, T.; Broeke, R.; Pleros, N. All-optical SR Flip-Flop based on SOA-MZI switches monolithically integrated on a generic InP platform. In Proceedings of the Smart Photonic and Optoelectronic Integrated Circuits XVIII, San Francisco, CA, USA, 23 May 2016.
34. Vagionas, C.; Fitsios, D.; Kanellos, G.T.; Pleros, N.; Miliou, A. All optical flip flop with two coupled travelling waveguide SOA-XGM switches. In Proceedings of the Conference on Lasers and Electro-Optics, San Jose, CA, USA, 6–11 May 2012.
35. Liu, L.; Kumar, R.; Huybrechts, K.; Spuesens, T.; Roelkens, G.; Geluk, E.J.; Vries, T.; Regrenny, P.; Thourhout, D.V.; Baets, R.; et al. An ultra-small, low-power, all-optical flip-flop memory on a silicon chip. *Nat. Photonics* **2010**, *4*, 182–187. [CrossRef]
36. Sakaguchi, J.; Katayam, T.; Kawaguchi, H. High Switching-Speed Operation of Optical Memory Based on Polarization Bistable Vertical-Cavity Surface-Emitting Laser. *IEEE J. Quantum Electron.* **2010**, *46*, 1526–1534. [CrossRef]
37. Pleros, N.; Apostolopoulos, D.; Petrantonakis, D.; Stamatiadis, C.; Avramopoulos, H. Optical static RAM cell. *IEEE Photonics Technol. Lett.* **2009**, *21*, 73–75. [CrossRef]
38. Vagionas, C.; Fitsios, D.; Kanellos, G.T.; Pleros, N.; Miliou, A. Optical RAM and Flip-Flops Using Bit-Input Wavelength Diversity and SOA-XGM Switches. *IEEE J. Lightwave Technol.* **2012**, *30*, 2012. [CrossRef]
39. Pitris, S.; Vagionas, C.; Tekin, T.; Broeke, R.; Kanellos, G.T.; Pleros, N. WDM-enabled Optical RAM at 5 Gb/s Using a Monolithic InP Flip-Flop Chip. *IEEE Photonics J.* **2016**, *8*, 1–7. [CrossRef]
40. Alexoudi, T.; Fitsios, D.; Bazin, A.; Monnier, P.; Raj, R.; Miliou, A.; Kanellos, G.T.; Pleros, N.; Rainieri, F. III-V-on-Si Photonic Crystal nanocavity laser technology for optical Static Random Access Memories (SRAMs). *IEEE J. Sel. Top. Quantum Electron.* **2016**, *22*, 1–10. [CrossRef]
41. Nozaki, K.; Shinya, A.; Matsuo, S.; Suzaki, Y.; Segawa, T.; Sato, T.; Kawaguchi, Y.; Takahasi, R.; Notomi, M. Ultralow-power all optical RAM based on nanocavities. *Nat. Photonics* **2012**, *6*, 248–252. [CrossRef]
42. Fitsios, D.; Vyrsokinos, K.; Miliou, A.; Pleros, N. Memory speed analysis of optical RAM and optical flip-flop circuits based on coupled SOA-MZI gates. *IEEE J. Sel. Top. Quantum Electron.* **2012**, *18*, 1006–1015. [CrossRef]
43. Vagionas, C.; Fitsios, D.; Vyrsokinos, K.; Kanellos, G.T.; Miliou, A.; Pleros, N. XPM- and XGM-based Optical RAM memories: Frequency and Time domain theoretical analysis. *IEEE J. Quantum Electron.* **2014**, *50*. [CrossRef]
44. Vagionas, C.; Bos, J.; Kanellos, G.T.; Pleros, N.; Miliou, A. *Efficient and Validated Time Domain Numerical Modelling of Semiconductor Optical Amplifiers (SOAs) and SOA-Based Circuits In Some Advanced Functionalities of Optical Amplifiers*; Intech Open Publishing: Rijeka, Croatia, 2015; pp. 1–26.
45. Vagionas, C.; Markou, S.; Dabos, G.; Alexoudi, T.; Tsiokos, D.; Miliou, A.; Pleros, N.; Kanellos, G.T. Column Address Selection in Optical RAMs With Positive and Negative Logic Row Access. *IEEE Photonics J.* **2013**, *5*. [CrossRef]
46. Alexoudi, T.; Papaioannou, S.; Kanellos, G.T.; Miliou, A.; Pleros, N. Optical cache memory peripheral circuitry: Row and column address selectors for optical static RAM banks. *IEEE J. Lightwave Technol.* **2013**, *31*, 4098–4110. [CrossRef]
47. Vagionas, C.; Pitris, S.; Mitsolidou, C.; Bos, J.; Maniotis, P.; Tsiokos, D.; Pleros, N. All-Optical Tag Comparison for Hit/Miss Decision in Optical Cache Memories. *IEEE Photonics Technol. Lett.* **2015**, *28*, 713–716. [CrossRef]
48. Maniotis, P.; Fitsios, D.; Kanellos, G.T.; Pleros, N. Optical Buffering for Chip Multiprocessors: A 16GHz Optical Cache Memory Architecture. *IEEE J. Lightwave Technol.* **2013**, *31*, 4175–4191. [CrossRef]
49. Apostolopoulos, D.; Zakynthinos, P.; Stampoulidis, E.; Kehayas, E.; McDougall, R.; Harmon, R.; Poustie, A.; Maxwell, G.; Caenegem, R.V.; Colle, D.; et al. Contention Resolution for Burst-Mode Traffic Using Integrated SOA-MZI Gate Arrays and Self-Resetting Optical Flip-Flops. *IEEE Photonics Technol. Lett.* **2008**, 2024–2026. [CrossRef]
50. Pitris, S.; Vagionas, C.; Maniotis, P.; Kanellos, G.T.; Pleros, N. An Optical Content Addressable Memory (CAM) Cell for Address Look-Up at 10Gb/s. *IEEE Photonics Technol. Lett.* **2016**, *28*, 1790–1793. [CrossRef]
51. Maniotis, P.; Terzenidis, N.; Pleros, N. Optical CAM architecture for address lookup at 10 Gbps. Proceeding of the SPIE Photonics West Optical Interconnects XVII, San Francisco, CA, USA, 28 January 2017.

52. Kehayas, E.; Vyrsokinos, K.; Stampoulidis, L.; Christodoulopoulos, K.; Vlachos, K.; Avramopoulos, H. ARTEMIS: 40-gb/s all-optical self-routing node and network architecture employing asynchronous bit and packet-level optical signal processing. *IEEE J. Lightwave Technol.* **2006**, *24*, 2967–2977. [CrossRef]

53. Pleumeekers, J.L.; Kauer, M.; Dreyer, K.; Burrus, C.; Dentai, A.G.; Shunk, S.; Leuthold, J.; Joyner, C.H. Acceleration of gain recovery in semiconductor optical amplifiers by optical injection near transparency wavelength. *IEEE Photonics Technol. Lett.* **2002**, *14*, 12–14. [CrossRef]

54. Junsangsri, P.; Lombardi, F.; Han, J. A Ternary Content Addressable Cell Using a Single Phase Change Memory (PCM). In Proceedings of the Great Lakes Symposium on VLSI, Pittsburgh, PA, USA, 20–22 May 2015.

55. Lengle, K.; Nguyen, T.N.; Gay, M.; Bramerie, L.; Simon, J.C.; Bazin, A.; Raineri, F.; Raj, R. Modulation contrast optimization for wavelength conversion of a 20 Gbit/s data signal in hybrid InP/SOI photonic crystal nanocavity. *Opt. Lett.* **2014**, *39*, 2298–2301. [CrossRef] [PubMed]

56. Lu, G.; Guo, C.; Li, Y.; Zhou, Z.; Yuan, T.; Wu, H.; Xiong, Y.; Gao, R.; Zhang, Y. Serverswitch: A programmable and high performance platform for data center networks. In Proceedings of the 8th USENIX Conference on Networked Systems Design and Implementation, Boston, MA, USA, 30 March–1 April 2011.

57. Gonzalez, J.; Orosa, L.; Azevedo, R. Architecting a computer with a full optical RAM. In Proceedings of the Electronics, Circuits and Systems (ICECS) International Conference, Monte Carlo, Monaco, 11–14 December 2016.

![applied sciences logo] *applied sciences*

MDPI

Article

Comparison of Basic Notch Filters for Semiconductor Optical Amplifier Pattern Effect Mitigation

Zoe V. Rizou [1],*, Kyriakos E. Zoiros [1] and Antonios Hatziefremidis [2]

[1] Democritus University of Thrace, Department of Electrical and Computer Engineering, Laboratory of Telecommunications Systems, Lightwave Communications Research Group, Xanthi 67 100, Greece; kzoiros@ee.duth.gr
[2] Technological Educational Institute of Chalkis, Department of Aircraft Technology, Chalkis 34 100, Greece; ahatzi@teihal.gr
* Correspondence: zrizou@ee.duth.gr; Tel.: +30-2541-079975

Received: 12 July 2017; Accepted: 28 July 2017; Published: 2 August 2017

Abstract: We conduct a thorough comparison of two basic notch filters employed to mitigate the pattern effect that manifests when semiconductor optical amplifiers (SOAs) serve linear amplification purposes. The filters are implemented using as the building architecture the optical delay interferometer (ODI) and the microring resonator (MRR). We formulate and follow a rational procedure, which involves identifying and applying the appropriate conditions for the filters' spectral response slope related to the SOA pattern effect suppression mechanism. We thus extract the values of the free spectral range and detuning of each filter, which allow one to equivocally realize the pursued comparison. We define suitable performance metrics and obtain simulation results for each filter. The quantitative comparison reveals that most employed metrics are better with the MRR than with the ODI. Although the difference in performance is small, it is sufficient to justify considering also using the MRR for the intended purpose. Finally, we concisely discuss practical implementation issues of these notch filters and further make a qualitative comparison between them in terms of their inherent advantages and disadvantages. This discussion reveals that each scheme has distinct features that render it appropriate for supporting SOA direct signal amplification applications with a suppressed pattern effect.

Keywords: semiconductor optical amplifier (SOA); pattern effect; optical delay interferometer (ODI); microring resonator (MRR)

1. Introduction

Optical amplifiers (OAs) are key elements for the development, implementation and evolution of fiber-based transportation, distribution and access networks [1,2]. Their traditional functions include (Figure 1 [3]) enhancing the signal power before being launched into an optical link (boosters), compensating for signal losses incurred either by the fiber medium, components along the propagation path or optical splitters, branches and taps (in-line reach extenders) and increasing the level of the received signal before photodetection (preamplifiers).

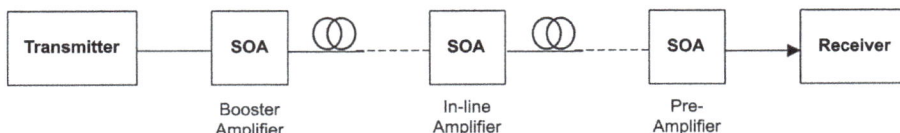

Figure 1. SOA direct amplification applications.

In this manner, they decisively contribute to achieving high global capacities, long transmission spans, multipoint-to-multipoint connectivity and ubiquitous information availability. These are critical requirements for efficiently coping with the growing data volumes and the diverse users' needs that govern the changeable broadband environment. Especially OAs that exploit semiconductor materials in the form of 'SOA' devices have been enjoying continuous preference by the optical communications research [4] and commercial (see indicatively [5]) sectors. The reason for this fact stems from SOAs' potential for signal amplification and multi-functionality entirely in the optical domain, within a broad wavelength range, with low power consumption, in a tiny volume at reasonable cost.

Despite SOAs attractive features, the employment of these elements for the aforementioned linear amplification purposes in diverse applications, such as optical transmission [6], radio over fiber [7], passive optical networks [8], optical wireless communications [9], optical data interconnects [10], converged metro-access networks with heterogeneous services [11] and burst information handling [12], has been compromised by SOAs carrier lifetime, which is finite [3]. When this physical parameter is comparable to the repetition period of the SOA driving data, as is often the case, it dominates the time scale of the SOA gain dynamics. Thus, if the combination of the power and duration of the input data signal is such that the SOA is deeply saturated, then the SOA gain is not perturbed and recovered in the same way for all excitation pulses (Figure 2).

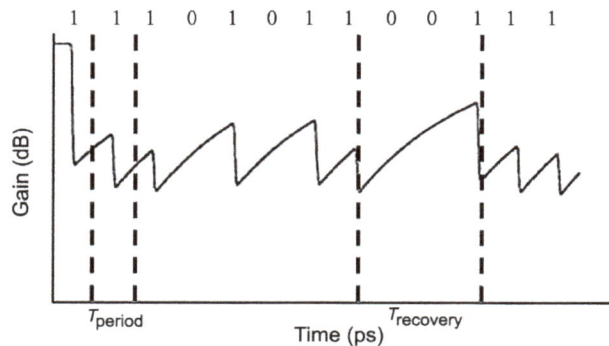

Figure 2. Evolution of semiconductor optical amplifier (SOA) gain in response to data train input of repetition period, T_{period}, such that pulses arrive faster than the interval available for SOA carrier replenishment, $T_{recovery}$.

As a consequence, the SOA operation becomes pattern dependent so that the amplification received by a given bit does not depend solely on this input, but also on the history of the SOA response to previous bits. This undesirable situation is referred to as the 'pattern effect' (PE) [13] and is more pronounced for data pulses that occupy a fraction of the assigned bit slot, i.e., of return-to-zero (RZ) data format, and hence, impose a heavy strain on the SOA gain dynamics [14]. Due to the PE, the profile of the pulses at the SOA output is not uniform, since pulses are distorted by peak-to-peak amplitude fluctuations, and therefore is far from ideally being an amplified replica of the input signal (Figure 3).

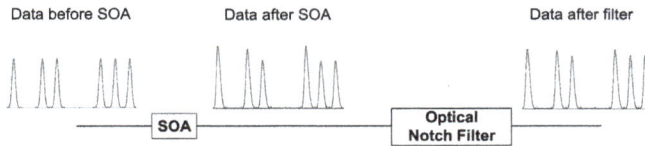

Figure 3. Data amplification in SOA with pattern-dependent output distortion and compensation by optical notch filtering.

In order to allow conventional SOAs to fulfil their classical role, it is necessary to combat the PE and its deleterious consequences. For RZ pulses, this requires lowering their power well below the SOA strong saturation regime, as well as substantially shortening their width [15]. These actions are not sufficient since the first one results in a poor optical signal-to-noise ratio (OSNR), reduced maximum output power, closer amplifier spacings and narrower input power dynamic range (IPDR) [16], while the second one requires applying intricate ultrafast pulse generation approaches [17] and dealing with practical related issues, such as nonlinear effects [18] and intraband phenomena [19].

Among the various methods that have been proposed to improve the pattern-dependent performance of SOAs used for direct signal amplification (see [15] and the relevant references therein), optical filtering has attracted intense research interest [20–25]. This passive method has been conceived of based on the observation that pulse amplification in an SOA is accompanied by spectral broadening to longer wavelengths (red shift) due to the manifestation of self-phase modulation (SPM) [26]. The SPM-induced spectral shift to longer wavelengths is higher than for the lower amplitude amplified pulses. This means that placing an optical filter after the SOA to pass a larger portion of the broadened spectrum for the less intense amplified pulses while blocking it for the more intense amplified pulses can compensate for the uneven red shift, which is converted then by the filter's slope [27] into more equalized pulse peak amplitudes (Figure 3).

During last few years, we have applied the optical filtering technique for SOA pattern effect suppression by means of a platform comprised of different notch filters [15,28–33]. Among these filters, we particularly distinguish the optical delay interferometer (ODI) [30–32] and the microring resonator (MRR) [33], which share several common features, such as an all-optical and passive nature, simple structure, compatibility both with fiber medium and microelectronic fabrication processes, compact size, periodic and tunable transfer function and potential for integration and co-packaging with the SOA in a single module. These filters have conventionally been destined to serve more classical filter-oriented applications. However, in this paper, we do not treat them individually, but in conjunction with the SOA and the serious PE problem of the latter, which they intend to solve. Since the MRR has recently received intense interest and is being widely exploited in numerous and diverse light wave applications, but the ODI has been employed more for assisting the SOA operation, it would be useful to investigate and assess whether this trend of shifting to the MRR technology should be extended to the mitigation of the SOA PE. To this aim, we conduct a quantitative and qualitative comparison of the ODI and MRR capability to confront the SOA pattern effect, thereby completing our work on the specific research topic. The comparison reveals that although the ODI suitability for the intended goal has been well tested and confirmed, both theoretically and experimentally, the MRR can realize better improvements in a set of key performance criteria and features several better operating characteristics than the ODI. This indicates that the MRR also has the technological potential to contribute in equal terms with its ODI counterpart for resolving the complications provoked by the SOA pattern effect.

2. Basic Optical Notch Filters' Configurations

The considered basic optical notch filters differ in their construction and operation and accordingly in the way that their spectral response is obtained and tailored for suppressing the SOA pattern effect.

The ODI's principle of operation relies on splitting, delaying and recombining the amplified signal, which is a process realized by means of intensity discrimination. Other waveguide-based technologies, which have been used for performing these functions in the wavelength domain, are reported in [34–38]. As shown in Figure 4a, it is constructed from two 3-dB couplers by interconnecting the output ports of the first one to the respective input ports of the second one. The upper arm of the formed interferometric configuration has a relative delay, $\Delta\tau$, against the second arm, into which a phase bias, $\Phi_{b,ODI}$, is introduced. When the amplified signal is inserted in the ODI from the first coupler, it is halved into two identical copies, which follow distinct paths along the ODI and acquire a wavelength-dependent phase difference analogous to their relative delay. Thus, when these components recombine at the second coupler, they interfere either constructively or destructively and produce at the ODI crossed output port the spectral response shown in Figure 5 with red color. The wavelength separation between adjacent maxima (peaks), or the free spectral range (FSR), is inversely proportional to the ODI relative delay, i.e., $FSR_{ODI} = \lambda_{amp}^2/(c\Delta\tau)$ [32], where λ_{amp} is the reference optical wavelength of the amplified signal, which lies in the vicinity of 1550 nm, and c is the speed of light in a vacuum. The minima (notches) are located halfway between maxima, and their exact wavelength position with respect to the shorter (blue) or longer (red) sideband, indicated by $\lambda_{notch,ODI}$ in Figure 5, depends on the phase bias, $\Phi_{b,ODI}$. This determines the wavelength detuning of the data carrier from the nearest transmission peak, $\Delta\lambda_{max,ODI}$, according to $\Phi_{b,ODI} = -2\pi \times (\Delta\lambda_{max,ODI}/FSR_{ODI})$ [39].

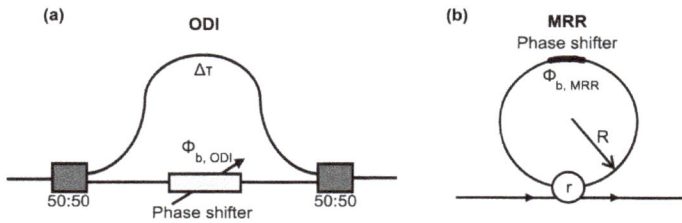

Figure 4. (**a**) Optical delay interferometer (ODI) and (**b**) microring resonator (MRR) structure.

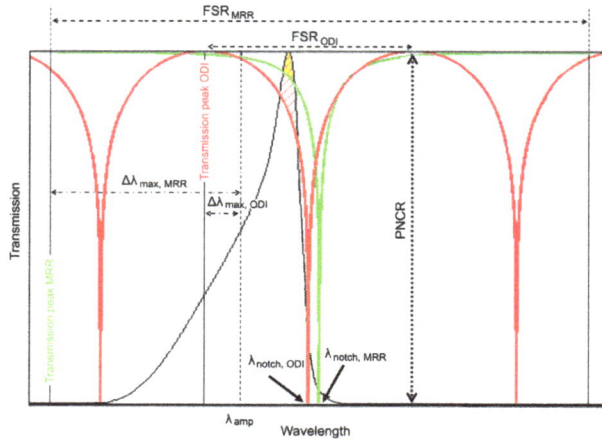

Figure 5. ODI (red color) and MRR (green color) spectral response together with spectrum at SOA output (black color). The red-colored oblique line and the yellow-colored zones show the red-shifted spectral components, which are filtered by the ODI and MRR, respectively.

The MRR is a waveguide shaped into a ring structure that is coupled to a bus waveguide, as shown in Figure 4b. When a signal enters this configuration, part of it is fed back to the MRR input while the rest is transferred at the MRR output. The exact signal magnitude and wavelength that penetrate into the ring depend on the strength of coupling between the straight and circular waveguides and whether the condition of resonance is satisfied. The amplified signal components that enter the ring subject to these conditions travel around it and after a time delay proportional to the ring circumference interfere with those components that pass directly to the exit. Then, provided that the MRR operates in the so-called critical coupling regime, where the field transmission coefficient, r, and the amplitude attenuation factor, l, are equal, the circulating intensity is maximized, and the transmitted intensity becomes minimum [33]. As a result, the spectral response shown with green color in Figure 5 is obtained. The notches, which are located at $\lambda_{notch,MRR}$, occur at resonance according to the condition $2\pi R n_{eff} = m\lambda_{notch,MRR}$, where R is the MRR radius, n_{eff} is the waveguide effective refractive index and m is a non-zero integer. The wavelength distance between consecutive maxima is given by $FSR_{MRR} = \lambda_{amp}^2/(2\pi R n_{eff})$. The spectral alignment of the transmission maximum, relative to the data carrier, $\Delta\lambda_{max,MRR}$, which, analogously to the ODI, corresponds to a phase bias $\Phi_{b,MRR} = -2\pi \times (\Delta\lambda_{max,MRR}/FSR_{MRR})$, can be varied within half FSR towards the blue or red sideband by tuning the MRR resonant wavelength through various appropriate mechanisms [40].

3. Optical Notch Filter Requirements for SOA Pattern Effect Suppression

The profiles of the ODI and MMR spectral responses in Figure 5 allow both schemes to act as optical notch filters for mitigating the SOA pattern-dependent performance degradation. For this purpose, it is necessary to suitably tailor the spectral characteristics of these responses so that the spectral components of the amplified pulses are forced to lie close to the notches and, hence, are suppressed in direct analogy to the degree of their red shift. This involves choosing and controlling the wavelength spacing (FSR), the contrast and the position of the notches, according to the following requirements [33]: (1) The FSR must be such that the notches are not spaced too close or too far, but sufficiently apart. Thus, making the FSR either too small or large prevents the red-shifted components of the amplified signal from distinctly falling along the response's slope and accordingly from being filtered analogously to their SPM-induced spectral magnitude. Instead, the FSR must be chosen so as to ensure that: (a) there is a sufficiently wide margin available for the red-shifted spectral components to be contained within the spectral borders defined by the difference $\lambda_{notch,ODI} - \lambda_{amp}$ and $\lambda_{notch,MRR} - \lambda_{amp}$; (b) the passband width is neither too narrow nor too broad so that the red-shifted spectral components do not fall outside or are loosely acted upon, respectively; (c) the suppression exerted by the filter on these components scales with the uneven peak amplitude they originate from, in order to clamp the higher amplitude amplified pulses while comparatively enhancing the lower amplitude ones; (d) the notches antipodes, i.e., the transmission peaks, are close enough so that the reference data wavelength can fall in their vicinity and perceive a sufficient fraction of the transmission maximum. In this manner, the amplified signal can pass through the notch filter without suffering a significant reduction in its amplitude or equivalently without significant elimination of the useful information it carries [21], which is critical when employing spectral elimination techniques for combating signal impairments [24]. (2) The repetitive notches must be sharp and deep enough to maximize their magnitude difference from their adjacent transmission peaks, which defines the peak-to-notch contrast ratio (PNCR) and ensure that the asymmetrically-broadened spectral components after the SOA are suppressed not weakly, but strongly and not identically, but to an extent directly dependent on the position in the data stream of the pulse from which they originate. (3) The notches must occur at a higher wavelength than that of the amplified signal, so that the transmittance is decreased as the wavelength is increased. This is again indispensable in order for the spectrally-broadened components to be suppressed analogously to the degree of their red shift.

4. ODI vs. MRR Comparison Rationale

In order to realize the comparison between the ODI and MRR, we must select and assign the proper values to the parameters that critically affect their operation, i.e., ($\Delta\tau$, $\Phi_{b,ODI}$) and (R, $\Phi_{b,MRR}$), respectively, so that this comparison is done on a fair basis. For this purpose a rational approach is adopted, which is followed after suitably combining the evidence available and knowledge that have been acquired, both theoretically and experimentally, on these types of modules when employed as notch filters. To this aim, the starting point is to equate the maximum slope of their spectral responses. Taking this as the comparison criterion is dictated by the SOA pattern effect suppression filtering mechanism, according to which the irregularly-shifted spectral components of the amplified pulses are converted by each filter's slope into amplitude changes, which counteract those incurred by the SOA ([27], cf. Figure 4.18). On the other hand, the maximum slope is related to the sensitivity of these filters to input signal perturbations, like those after the SOA, and accordingly to their capability to handle and compensate them [41]. Then, the maximum slope can be determined by finding the zeros of the second derivative of each filter's transfer function and replacing them into the first derivative. This procedure, which is conveniently executed in the angular frequency domain, leads to the equality:

$$\frac{3\sqrt{3} \cdot r}{8(1 - r^2)} T_R = \frac{\Delta\tau}{2} \tag{1}$$

where the left-hand side gives the maximum slope at critical coupling of the MRR ([42], cf. Equations (3.11)–(3.12) and (B.9)), whose round-trip delay is $T_R = 2\pi n_{eff} R/c$, while the right-hand side represents the same quantity for the ODI [15]. Letting then $r = l = 0.95$, so as to ensure that the condition of critical coupling for the MRR is favorably met, but also to preserve the continuity with previous relevant work [33], leads to:

$$\Delta\tau \cong 12.6 T_R \tag{2}$$

which readily gives $T_R = 0.8$ ps, or equivalently, $R \cong 28$ um, for $\Delta\tau = 10$ ps, where the latter value has not been arbitrarily chosen, but according to the selection rules derived in [31] for the conditions under which the ODI should operate to combat the SOA pattern effect. Similarly, the calculated MRR radius complies with the permissible range of values specified for this parameter from the study in [33]. Furthermore, $\Delta\tau$ and R are such that the filtered-out data pulses retain their initial shape, as addressed in [15,29] and verified again below. Having determined the FSR of the ODI and MRR, the next step is to do the same for their detuning. This can be done by applying the criterion of equalization of the red-shifted components of the amplified signal, which must hold in order for the low-pass frequency characteristic of the SOA to be compensated by the high-pass frequency characteristic of the two employed notch filters ([27], cf. Figure 4.4). In this context, the filters' intensity response, $|H(\lambda)|^2_{ODI,MRR}$, is Taylor-expanded to the first order around λ_{amp}, thus leading to the following requirement in the linear scale for the filters' slope magnitude, K, [43], which governs the efficiency of converting the uneven SPM-induced spectral broadening to uniform amplified pulse amplitude [44]:

$$K_{ODI,MRR} = \left.\frac{d|H(\lambda)|^2_{ODI,MRR}}{d\lambda}\right|_{\lambda=\lambda_{amp}} = \frac{2T_{car}}{\alpha_{LEF}} \frac{\pi c}{\lambda^2_{amp}}. \tag{3}$$

This relationship essentially expresses in mathematical terms the fact that the SOA temporal response, which is limited by the SOA finite carrier lifetime, T_{car}, and the associated pattern effect, can be enhanced by means of the employed filters by a factor that scales at most with the SOA linewidth enhancement factor, α_{LEF} [45]. Then, using the electric field responses [40,46]:

$$H_{ODI}(\lambda) = \frac{1}{2}\left\{1 + \exp\left[-j\left(2\pi c\frac{\lambda - \lambda_{amp}}{\lambda_{amp}^2}\Delta\tau + \Phi_{b,ODI}\right)\right]\right\} \tag{4a}$$

$$H_{MRR}(\lambda) = \frac{r - l\exp\left\{-j\left[2\pi c\dfrac{\lambda - \lambda_{amp}}{\lambda_{amp}^2}T_R + (\pi + \Phi_{b,MRR})\right]\right\}}{1 - rl\exp\left\{-j\left[2\pi c\dfrac{\lambda - \lambda_{amp}}{\lambda_{amp}^2}T_R + (\pi + \Phi_{b,MRR})\right]\right\}} \tag{4b}$$

for the ODI and MRR, respectively, and substituting in (3) yields, after appropriate algebraic manipulations and numerical calculations for $T_{car} = 75$ ps and $\alpha_{LEF} = 8$, $\Phi_{b,ODI} \simeq 0.7\pi$ and $\Phi_{b,MRR} \simeq 0.98\pi$. Note that the MRR phase bias has mathematically been shifted by 'π', so that the transfer functions of both considered filters become null when the wavelength position of the input data carrier coincides with that of the nearest notch, which happens when $\Phi_{b,ODI} = \Phi_{b,MRR} = \pi$. These phase bias values correspond to $\Delta\lambda_{max,ODI} = -0.27$ nm and $\Delta\lambda_{max,MRR} = -0.48$ nm. The difference in the detuning values is attributed to the fact that, as shown in Figure 5, the ODI and MRR have different FSR, i.e., $FSR_{MRR} \simeq 10 \cdot FSR_{ODI}$. Furthermore, they imply that $\lambda_{notch,ODI}$ is closer to λ_{amp} than $\lambda_{notch,MRR}$, since $\lambda_{notch,ODI} - \lambda_{amp} = 0.25$ nm while $\lambda_{notch,MRR} - \lambda_{amp} = 0.27$ nm.

The suitability of the critical parameters values specified above is further verified by observing that both filters extended the modulation response bandwidth of the SOA, which is inherently limited by the SOA finite carrier lifetime responsible for the pattern effects. This is shown in Figure 6, which has been obtained by plotting first the modulation response of the stand-alone SOA and then combining it in a multiplicative manner with that of the ODI or MRR. These responses are mathematically described by closed-form expressions obtained through appropriate small-signal analysis, which allows one to find for each module the transfer function between small signal changes of the output power incurred by small-signal perturbations of the input power, as reported in [47] for the SOA and ODI and in [48] for the MRR.

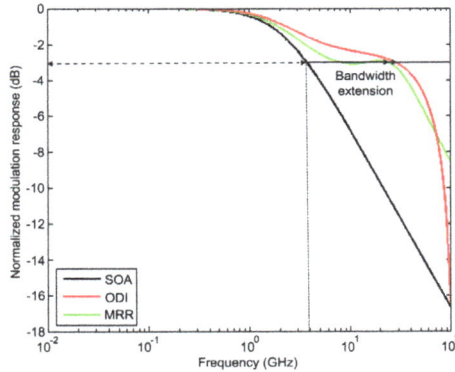

Figure 6. Small-signal modulation response of SOA alone (black color) and with the assistance of ODI (red color) or MRR (green color). The dashed line denotes the 3-dB bandwidth level and the arrows the bandwidth extension achieved by the notch filters.

5. ODI vs. MRR Comparison: Results and Discussion

The ODI and MRR are compared using appropriate performance metrics, which allow one to assess the performance of an SOA subject to a strong pattern effect and subsequently the capability of the employed notch filtering schemes to suppress it. These metrics include: (a) The amplitude modulation (AM), which is defined as AM (dB) $= 10 \log \left(P_{1,high} / P_{1,low} \right)$, where $P_{1,high}$ ($P_{1,low}$) is the peak power of the mark across the amplified data stream, which has the highest (lowest) amplitude level [30]. The AM provides a measure of the degree of uniformity of the amplified pulses whose pattern-dependent pulse-to-pulse wandering should be reduced by the employed filters below 1 dB [49]. (b) The amplification reduction (AR), which is defined as AR (dB) $= \left| P_{out/ODI,MRR}^{avg}(\text{dBm}) - P_{out/SOA}^{avg}(\text{dBm}) \right|$, where $P_{out/ODI,MRR}^{avg}$ and $P_{out/SOA}^{avg}$ are the total average powers at the ODI or MRR filters and SOA outputs, respectively: The AR takes into account the fact that, due to the suppression of the red-shifted components by the action of filtering, some useful information contained in them is inevitably lost [21]. This causes the data sequence to receive less amplification than after the SOA alone, which is translated into an amplification penalty [50] that is quantified by the AR. Ideally, the extent of AR must be such that it can be compensated by the average gain offered by the SOA to the data stream and accordingly allow the net gain of the SOA-notch filter combination to be sufficient for direct signal amplification purposes. (c) The cross-correlation (XC) coefficient, which is defined as [51],

$$XC \ (\%) = \int_{-\infty}^{\infty} \hat{P}_{out/ODI,MRR}(t)\hat{P}_{in/SOA}(t)\,dt \Bigg/ \sqrt{\left(\int_{-\infty}^{\infty} \hat{P}_{out/ODI,MRR}^{2}(t)\,dt \right) \left(\int_{-\infty}^{\infty} \hat{P}_{in/SOA}^{2}(t)\,dt \right)} \ \text{where}$$

the symbol of the caret over the time-dependent powers denotes that the latter are normalized to their maximum value: The XC is a measure of the degree of similarity of the intensity profile of the amplified signal that is filtered by the ODI or MRR to its profile prior to being inserted in the SOA. The higher the XC, the better is the capability of the given filter to accurately reproduce the original signal. (d) The Q^2-factor represents the OSNR at the input of the receiver decision circuit: in the thermal noise limit [52], where the amplitude fluctuations due to the pattern effect act as noise variance on the marks [19], $Q = P_{avg}^{1} / \sigma_{avg}^{1}$, where P_{avg}^{1} is the average and σ_{avg}^{1} is the standard deviation of the peak power of the marks and OSNR $= Q^2$. Because the OSNR is measured in units of decibels (dB), it is convenient to use the same dimensioning for the Q-factor. Thus, $10 \log(\text{OSNR}) = 10 \log(Q^2) = 20 \log Q \equiv Q^2$ (dB). On the other hand, the Q-factor value in the linear scale must be at least six to ensure that the associated bit error rate is less than 10^{-9} [52] and error-free operation can be achieved. Converting this requirement into logarithmic scale results in Q^2 (dB) $= 20 \log Q = 20 \log(6) \simeq 15.6$ dB, which therefore is the limit against which the specific employed metric is addressed in the rest of the paper. Then, it is possible to determine: (i) the input power dynamic range (IPDR) [53]; (ii) the permissible variation from the specified detuning; and (iii) the maximum signal transmission distance in standard single-mode fiber (SSMF) without dispersion compensation. The signal powers involved in the above definitions can be found through simulation run at 10 Gb/s. This bit rate is in line with data rates of modern applications that exploit SOAs as linear amplification elements [2,8] and falls in the range within which significant bit pattern distortions are caused on signals directly amplified by typical SOAs [54]. The input to the SOA is a pseudo random binary sequence (PRBS) of 127 bits in length, which allows one to fully capture and investigate the detrimental impact of the SOA pattern effect [55]. This data stream contains Gaussian-shaped RZ pulses of such duty cycle and peak power that together saturate heavily the SOA and provoke a strong pattern effect at its output [31]. The default parameters values used in the simulation are those reported in [31,33]. The simulation procedure involves [15,33] numerically calculating the SOA integrated gain temporal response, substituting the outcome in the electric field (normalized so that its squared modulus represents power [26]) of the optical signal at the SOA output, applying fast Fourier transform (FFT) to pass it into the frequency domain, convolving it with the ODI or MRR spectral response given from (4) and finally converting the convolution product back into the time domain using inverse FFT (IFFT). The FFT and IFFT operations are available and executed in MATLAB software. The calculation of the Q^2-factor is done according to the method described in [15]. The integrals involved in the definition of XC are numerically calculated in MATLAB using Simpson's rule.

Figures 7–11 show the simulation results obtained for each notch filter. To facilitate comparison, the waveforms and curves have been placed in the same figure with differently colored lines, i.e., red for the ODI and green for the MRR. From these graphical and numerical results, the following concise comments can be made with regard to each performance metric and accordingly to the SOA pattern effect suppression capability of the ODI and MRR:

(a) *AM*: Both filters can improve this metric and render it acceptable even between the '1' that follows after the longest string of repeated '0's, i.e., six in our case [56], and the '1' that appears just before this run, but also after the same number of consecutive '1's (Figure 7a). In fact, the *AM* between the target marks is reduced from 1.15 dB after the SOA (Figure 7b) to 0.22 dB by the ODI (Figure 7c) and to 0.34 dB by the MRR (Figure 7d), which therefore indicates that this demanding situation can equally be confronted with success. The stronger suppression of the *AM* achieved by the ODI is justified by the fact that, for the specified detuning values, the notch lies closer to the reference optical wavelength than for the MRR. Consequently, the ODI filters a greater portion of the broadened spectrum of the amplified signal, as indicated in Figure 5 by the red-colored oblique-line zone versus the yellow-colored zone for the MRR. This enhances the amplitude equalization of the spectral components that have asymmetrically been shifted to the longer sideband, but as explained in the following, this enhancement is compromised by the higher *AR* and the accompanying performance implications. Although the difference in the reduction of the *AM* seems to be small between the two filters, it is not insignificant. In fact, it may become important when cascading several SOAs [12,57], where the peak amplitude differences between amplified pulses should be prevented from being accumulated from stage to stage so that they do not become detrimental for the optical transmission performance.

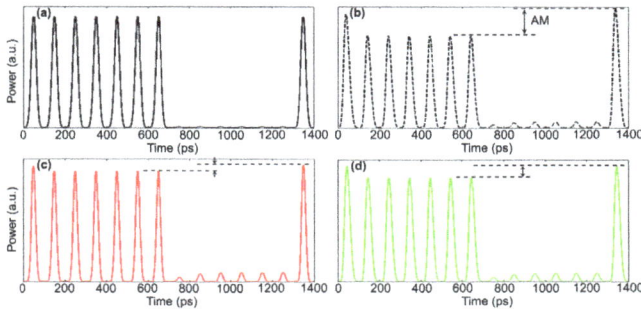

Figure 7. Simulated temporal waveforms that contain maximum strings of '1's and '0's. (**a**) SOA input, (**b**) SOA output, (**c**) ODI output and (**d**) MRR output. The vertical arrows indicate the amplitude modulation (*AM*).

(b) *AR*: Although the ODI filters more spectral components, as explained in (a), this also means that a greater portion of useful information contained in these components is inevitably lost. As a by-product, the data sequence receives less amplification than after the SOA alone, which is translated into an analogous difference between the *AR* of the two filters, i.e., 3 dB for the ODI and 2.78 dB for the MRR. This in turn affects the net gain of the SOA-notch filter system, which is further lowered by the losses inserted by each type of filter [15]. Although the loss per length unit of the MRR and the ODI is comparable [58,59], taking into account the significantly longer propagation length of the ODI, which is required in order to create the necessary length imbalance between its two arms, it can be deduced that the total losses exceed by a magnitude order of 3 dB those of the MRR, which are extremely low owing to the much smaller size of the latter. Thus, for the ODI, it is necessary to compensate for its increased losses by adjusting the SOA to provide more gain, which, as explained in [15], is an action that does not impair the capability of this notch filter to suppress the *AM*. This extra gain can be offered by supplying more carriers via a higher bias current [8], but at the inevitable expense of increased power consumption. Accordingly, if insertion losses (IL) are included, the MRR allows the SOA-filter system to exhibit a higher net gain [15], i.e., 12.13 dB vs. 8.91 dB for the ODI.

(c) *XC*: The numerically-obtained values, 97% for the MRR and 93% for the ODI, indicate that in both cases, the filtered pulse temporal waveform deviates only by 3% and 7%, respectively, from that at the SOA input. This degree of matching implies that the Gaussian-like power profile of the input data pulses is strongly preserved at the filters' output. This is visually confirmed in Figure 8a, which shows that the envelope of an isolated data pulse maintains its original form. However, the time-bandwidth product at the output of both filters is over the minimum defined for Gaussian-shaped pulses [52], which means that the filtered pulses become chirped. The profile of this chirp, which is shown in Figure 8b, reflects the qualitative differences in the data pulse profiles of Figure 8a. More specifically, in the pulse rising part (region of 10–35 ps), the chirp declines to negative values, i.e., red chirp, but this drop is more dominant for the ODI than for the MRR. This partly explains why the AMattained with the former notch filter is lower than that achieved with the latter. As the pulse continues to temporally evolve, the chirp is decreased more strongly for the ODI until we reach the pulse center, where its absolute magnitude becomes maximum. From this point onwards the situation is reversed and the chirp begins to incline with a slope that is sharper for the ODI than for the MRR. Consequently, and as explained in detail below, the chirp parameter is larger for the ODI, which has a negative impact on the transmission performance of this notch filter in the presence of fiber dispersion. Finally, as we move off the pulse center and approach the pulse falling part (region of 65–80 ps), the chirp changes sign and becomes positive, i.e., blue chirp, which, nevertheless, because of its different cause and time scale over its red chirp counterpart, is not of major concern with regard to the SOA pattern effect suppression [31].

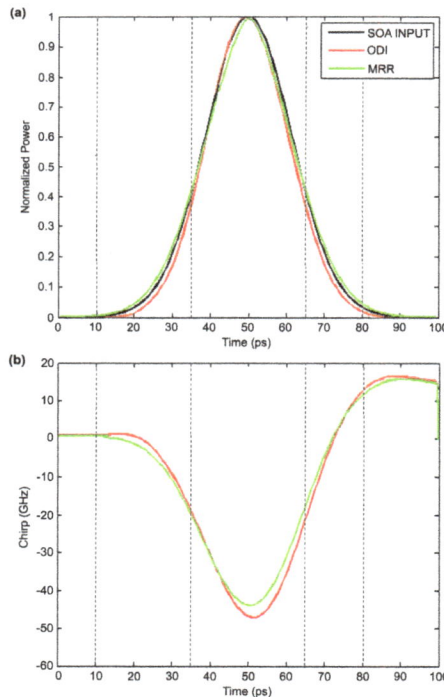

Figure 8. (**a**) Profile of isolated data pulse at ODI (red color) and MRR (green color) output in comparison to SOA input (black color). (**b**) Corresponding chirp profile at ODI (red color) and MRR (green color).

(d) Q^2-factor and associated metrics: (i) both filters enhance the IPDR and extend the SOA error- and distortion-free operation from the low saturation region (A in Figure 9) to the deep saturation region (C in Figure 9), in which the SOA gain is decreased by more than 3 dB and 6 dB, respectively. This fact allows one to passively relax the stringent management [60,61] of the input power levels going into the SOA and hence of the degree of SOA gain saturation, which otherwise is required in order to allow SOAs to deliver unimpaired linear amplification functionality. This improvement is realized for both filters, as quantified by the increase of the peak input data power (P_{peak}) by $\Delta P = 14$ dB (assuming a maximum power of 10 dBm going into the SOA). Furthermore, at $P_{peak} \simeq 6.6$ dBm, which heavily saturates the SOA [31], the Q^2-factor is over its critical limit, which for the SOA alone fell short by $\Delta Q^2 = 4.3$ dB. This is better achieved for the MRR than for the ODI, i.e., $\Delta Q^2 = 2.87$ dB and $\Delta Q^2 = 1$ dB, respectively.

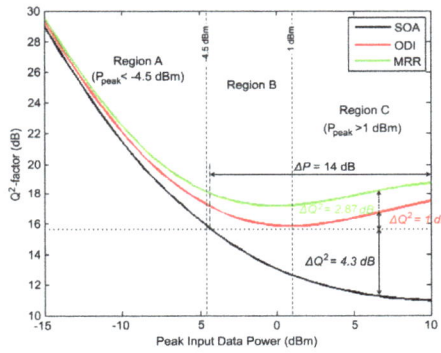

Figure 9. Q^2-factor as a function of different peak input data powers for SOA alone (black color) and with the connection of ODI (red color) and MRR (green color). The horizontal dotted line denotes the lower limit set for the Q^2-factor. The vertical dashed lines define the different saturation regions: A, low; B, medium; C, deep.

(ii) The MRR exhibits better tolerance to detuning offset than the ODI, which is an important feature as it obviates the need for resorting to electronic equalization [44]. In fact, Figure 10 shows that the (negative) detuning of the MRR can be varied within 0.12 nm or equivalently by 50% further for the data pulses wavelength modulation bandwidth [52] in the C-band and still keep the Q^2-factor acceptable. The corresponding variation for the ODI is comparatively reduced by more than 50%, i.e., 0.05 nm, and hence is narrower, which may impose tighter requirements with regard to wavelength locking [62].

(iii) The MRR allows the amplified signal to propagate over a longer distance of SSMF without dispersion compensation than the ODI. This is shown in Figure 11, which has been obtained by modeling the SSMF of typical attenuation coefficient 0.2 dB/km and dispersion parameter 17 ps/nm/km by its low-pass equivalent field transfer function [63]. From this figure, it can be seen first that subject to chromatic dispersion, the maximum SSMF length is reduced to around 20 km, as calculated from [64] (cf. Equation (1)) for intensity-modulated data pulses of a 100-ps repetition period and 27% duty cycle. Then, looking for the transmission range for which the Q^2-factor is acceptable, we observe that the amplified signal can travel up to 12 km with the ODI, while this span is extended to 15 km with the MRR. This length difference may seem small, but nevertheless, it corresponds to the reach of real short-haul fiber interconnection links and networks [65]. Moreover, the Q^2-factor is well over its lower permissible bound for the MRR, while it is much closer to the borderline for the ODI, across the SSMF distance over which this metric remains acceptable for both filters. These observations can be interpreted by referring to the instantaneous frequency deviation, i.e., chirp, which inevitably accompanies a signal when directly amplified in an

SOA [26]. More specifically, suppressing the SOA pattern effect by means of optical notch filtering relies on compensating for the chirp components, which are induced due to the irregular SOA gain variation by the optical excitation, and converting them into amplitude changes that counteract those after the SOA ([27], cf. Figure 4.18). Still, it is not possible to fully eliminate the SPM-induced negative-magnitude chirp components, which are directly associated with the SOA pattern effect, but only reduce the variations of their uneven peaks [31]. By calculating the maximum chirp variation, which occurs in the transition between the last '0' in a run of spaces and the '1' that immediately follows, we can extract the chirp parameter at the filters' output similarly to the numerical procedure described in [66]. The value of this parameter is positive and is larger for the ODI than for the MRR. In the presence of anomalous chromatic dispersion, this fact subsequently affects the pulse width of the chirped signal at the filters' output [52]. Thus, after a short distance zone of pulse compression, pulse broadening occurs, which is more significant for the ODI than for the MRR. This is reflected in the Q^2-factor, which is smaller for the ODI and becomes worsened after propagation in a shorter SSMF length (which is also smaller than that calculated from [64] (cf. Equation (1)) for the chirped-free signal case). The obtained results are further supported by the so-called pseudo-eye diagrams (PEDs) [67], which are displayed at different points of the Q^2-factor curves. Thus, both MRR and ODI restore the form of the PED, which has severely been degenerated into asymmetric sub-envelopes at the SOA output. This improvement is further quantified by the degree of eye opening (EOP) [68], which is higher for the MRR than for the ODI, i.e., 91% vs. 87%, respectively, for transmission in the same fiber length of 12 km.

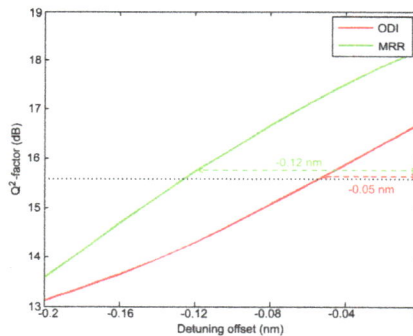

Figure 10. Q^2-factor for different detuning offset values for ODI (red color) and MRR (green color). The zero detuning refers to the case that the notch of the filters is located away from the data carrier by so many spectral units as defined in Figure 5.

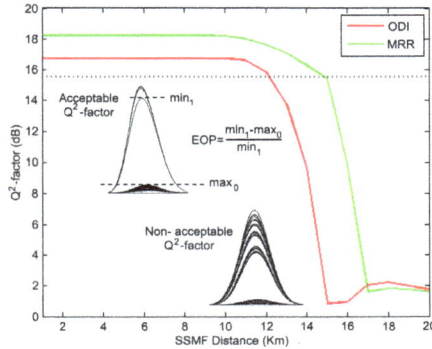

Figure 11. Q^2-factor vs. standard single-mode fiber (SSMF) distance for ODI (red color) and MRR (green color). The horizontal dotted line denotes the lower limit set for the Q^2-factor.

The above improvements can be attributed to the different shape of the MRR and ODI transfer functions [69]. This difference affects the efficiency of the frequency-to-amplitude conversion mechanism, which is essential for suppressing the SOA pattern effect [44] and thus makes the MRR and ODI perform differently when employed as notch filters. Moreover, the comparatively better performance of the MRR can be justified from a qualitative standpoint by noting that the MRR exhibits higher wavelength sensitivity than the ODI. In fact, for the MRR, this quantity is determined, and hence enhanced, by the finesse [70], which is a function of the coupling coefficient and internal loss [27] and accordingly can be flexibly adjusted. In contrast, the finesse of the ODI is very small, i.e., it equals only two [52], and therefore is fixed so that the design freedom is very limited. Physically, this happens due to the different operating nature of the ODI and MRR, since in the former, only two constituents of the input field interfere to produce an output, whilst in the MRR, the field experiences multiple circulations before it leaves the resonant cavity [71]. Since in the first case, the interfering paths are feeding forward, while in the second case there is a recirculating, feedback delay path, this difference, which is reflected on the polynomial form of the respective transfer functions, can also be expressed in signal processing terms. Accordingly, the ODI and MRR can be used not only as notch filters, but in general as finite impulse response (FIR) and infinite impulse response (IIR) filters, respectively [72]. FIR filters that exploit MRRs are also possible if sophisticated designs based on the combination of multiple such elements are adopted [73].

The values obtained for the employed performance metrics have been gathered in Table 1.

Table 1. Values of employed performance metrics.

Parameter	ODI	MRR
AM (dB)	0.22	0.34
AR (dB)	3	2.78
Net gain (dB)	8.91 *	12.13
XC (%)	93	97
ΔP (dB)	14	14
ΔQ^2 (dB)	1	2.87
(negative) Detuning offset (nm)	0.05	0.12
SSMF distance (km) w/o dispersion compensation	12	15
EOP at 12 km (%)	87	91

* 3 dB insertion losses included.

6. ODI vs. MRR Qualitative Comparison

The key characteristics that the ODI- and MRR-based notch filters must exhibit in order to mitigate the SOA pattern effect can be further assessed in a qualitative manner by considering the following functional and technological issues:

(a) FSR:

The FSR of both filters can conveniently lie within the SOA wide gain spectrum enabled through suitable adjustment of the SOA active layer material composition [1]. For the ODI, the FSR can be adjusted either statically, by fixing the path length imbalance, which can be of the order of um, between the ODI upper and lower arms, or dynamically, by using an optical delay line [32]. In the first case, a precise construction is required, but which is compensated by the reduction in footprint [74], while in the second case, a wide range of relative delays can be offered with fine resolution at the expense of a more bulky configuration [29]. For the MRR, the FSR can be readily set by choosing the MRR radius, which can be down to the micrometer scale using high index contrast waveguide materials [40,75]. This offers the comparative benefit of ultra-compact total size at the expense of resorting to elaborate and delicate microelectronic fabrication processes, which detracts from the global use of the specific scheme.

(b) Detuning:

The position of the notches relative to the input data carrier wavelength can be practically adjusted by exploiting effects such as thermo-optic. For this purpose, either active phase shifters are used to control the phase bias in the ODI [58] or electrical heaters in the direct vicinity of the MRR to spectrally shift its resonance [75]. This method is particularly efficient for materials of large positive thermo-optic coefficients, such as silicon, since the required amount of detuning across the FSR can be achieved with a power supply of the order of mWs for both filters. On the other hand, the need for precise detuning is inherently more demanding for the ODI than for the MRR. This happens due to the different way by which the amount of phase shift required for efficiently suppressing the SOA pattern effect is obtained. Thus, for the ODI, this amount is acquired through one-way interference of a pair of delayed signal copies, while for the MRR through periodic round-trips and resonant enhancement. Accordingly, the ODI operation is more sensitive to deviations from the condition of destructive interference, which must be tightly satisfied at its cross-wise output port [30], thus requiring special design [76,77].

(c) PNCR:

For the ODI the optimum PNCR can be achieved by adjusting the power splitting ratios of the optical couplers that form its two branches to be as symmetric as possible, i.e., as close as possible to 3 dB [78]. In practice, in order to ensure a PNCR over 20 dB, the deviation from the perfect 3-dB splitting ratio must be kept within 10% [29]. This tolerance is satisfiable with the coupling technology employed both in the fiber [79] and the waveguide [80] versions of the ODI.

Similarly, for the MRR, it is necessary to satisfy the condition of critical coupling, where the MRR field transmission coefficient, r, and amplitude attenuation factor, l, are equal while tending to unity [40]. In effect, the PNCR exceeds 20 dB even if this matching is less than 3% perfect [81]. The required matching can be accomplished by means of electrically-driven micro-electromechanical system (MEMS) microactuators, which control the gap by deforming the straight waveguide against the ring [82]. The ring, waveguide and MEMS actuators can be patterned with standard optical lithography, thus rendering the fabrication of the whole system technologically feasible.

(d) Temperature sensitivity:

Although the strong thermo-optic effect of materials used as a platform for constructing the considered filters allows one to efficiently tune them, it is also responsible for making them prone to temperature fluctuations [83]. This susceptibility may cause the bias point of the ODI and the resonance wavelength of the MRR to drift, which in turn could affect the process of suppressing the red-shifted spectral components without significantly impairing the input data carrier and eventually leading to temperature-dependent performance degradation. Although it was numerically verified above that the MRR is, by its principle of operation, more tolerant to detuning offsets than the ODI, still, special care must additionally be taken in order to enhance the resilience of both schemes to possible temperature changes while at the same time achieving this goal in an energy-efficient manner. To this aim, it has been identified [83] that while athermal operation of the ODI is possible without any extra energy consumption, there is no efficient passive solution to the problem for the MRR, as well. Furthermore, the exploitation of negative thermo-optic materials that could resolve the thermo-optic issue in a straightforward manner faces challenges with regard to compatibility with standard CMOS technology and is the subject of intense research.

(e)   Integration potential:

Both schemes are amenable to integration, which is highly desirable for keeping size and IL small while enabling interconnection even at a small scale level. Thus, both are compatible with planar light wave technology and can be co-packaged with the SOA in the same module in a monolithic (ODI) [80] or hybrid (MRR) [67,84] platform. Especially, the capability for co-packaging is rather crucial towards compensating for the reduction in the amplified signal level due to notch filtering and hence for extending transmission reach. Since the MRR can be made more compact than the ODI [83,85], the use of the former could be favored for mitigating the SOA pattern effect in applications where space availability is critical, as in the transmitter side for power boosting, especially of multiple wavelength outputs, or the receiver end for power preamplification, or within intra-data center optical links [85]. On the other hand, it should be noted that both filters are compatible with standard CMOS (complementary metal-oxide-semiconductor) processing in silicon photonics platform [85,86] and hence can leverage the benefits that accompany this mature fabrication technology.

(f)   Multi-wavelength operation:

Both filters can mitigate the SOA pattern effect induced on wavelength-division-multiplexed (WDM) data inputs, for all channels simultaneously [87]. The specific potential is enabled by the periodic spectral response of these filters and can be exploited provided that the center wavelengths of the different input data are spaced apart by integer multiples of the FSR [21,58]. In this case, the amplitude equalization can be realized by means of a single such optical notch filter, which is attractive from an economical point of view, since the cost of using the device can be shared among the multiple served data inputs and remain affordable with the increase of channel count [62].

Furthermore, the two types of filters can be discriminated, with regard to their efficiency, to enhance the amplitude uniformity of multiple SOA amplified data-carrying channels, by the amount of the achieved FSR, which is compared against the wavelength separation between adjacent channels in real deployed WDM communication systems. Thus, for the ODI, the FSR that corresponds to a relative delay of $\sim$10 ps between its two arms is almost the standardized dense WDM grid spacing of 0.8 nm at around a 1550-nm carrier wavelength. On the contrary, the MRR μm radial size results in an FSR that is an order of magnitude higher and accordingly more suitable for coarse WDM purposes. Another option is to cascade several MRRs of different radii along the same waveguide bus [88]. This allows one to accommodate more data wavelengths impaired by the SOA pattern effect and with a more dense spacing than for a single MRR [33], at the expense of a larger footprint and a more complex configuration. On the other hand, in WDM systems of some given interchannel spacing and spectral width, a single MRR can support numerous channels in a much more flexible manner than a single ODI. In fact, this capability critically depends on one filter's finesse, which, as previously mentioned, can be flexibly adjusted for the MRR, whereas it is very low and fixed for the ODI. Furthermore, the sharper spectral selectivity of the MRR allows the latter to better handle variations in the channel positions and compensate for possible misalignments against notches. This further allows one to use uncooled, i.e., non-temperature-controlled, single-mode lasers, which is an attractive feature from a cost and operational perspective [8]. In contrast, these functional requirements can be met with the ODI only if multiple ODI stages are concatenated [52], which, along with the aforementioned inevitable compromises associated with the employment of chains of filters, comes at the additional cost of extra losses.

(g)   Tunability:

The tunability of each type of filter can be assessed against bandwidth tunability and wavelength tunability [89]. For the first ability, we can calculate its minimum, which can be done in a straightforward manner by using the analytic formulas that quantify it. These are given from [52] $FWHM_{ODI} = FSR_{ODI}/2$ for the ODI and $FWHM_{MRR} = \lambda_{amp}^2(1-rl)/(2\pi^2 n_{eff} R \sqrt{rl})$ [75]. Recall now that in order to prevent the filtering action that suppresses the SOA pattern effect on the data pulses from distorting the latter, the ODI delay should not exceed 10 ps [31], while the MRR radius should not be larger than 42 um [33]. Then, substituting these maximum values together with $\lambda_{amp} = 1550$ nm, $r = l = 0.95$ and $n_{eff} = 1.41$ in the above expressions of the FWHM, it can be found that $FWHM_{ODI} \geq 0.4$ nm and $FWHM_{MRR} \geq 21$ nm. For the second ability, on the other hand, we have shown from Figure 10 that the wavelength offset of the MRR is allowed to be larger than that of the ODI while maintaining an acceptable Q^2-factor. Thus, in both cases, the numerically-obtained numbers suggest that the MRR filter is better tunable for the target application of SOA pattern effect compensation.

Table 2 summarizes the qualitative comparison of the MRR and ODI key characteristics. Note that this comparison is not absolute in the sense that each filter has its own special features, which may render it more suitable than the other for exploitation in SOA direct amplification applications with a suppressed pattern effect.

Table 2. Qualitative comparison of MRR and ODI key characteristics.

	Characteristic	MRR	ODI
	Width	Coarse	Dense
	Adjustment	Rather fine (radius)	Fine (path length imbalance) or bulky (optical delay line)
Free Spectral Range (FSR)	Multi-wavelength Operation	Yes	Yes
	(i) ITU -Grid	Difficult for DWDM since it requires cascading; is compromised by increased footprint and complexity	Easy for DWDM
	(ii) fixed interchannel spacing and spectral width	Yes, easy with single MRR owing to high and controllable finesse	Yes, difficult due to low and fixed finesse
	(iii) handling of variations in channel positions	Yes, straightforward owing to sharp spectral selectivity	Yes, requires to concatenate multiple stages at the cost of increased complexity and insertion losses
Detuning	Feasibility	Yes, thermo-optic effect with mWs/FSR	
	Mechanism	Relaxed (periodic round-trips and resonant enhancement)	Tight (one-way interference)
	Precision	Normal	Demanding
	Tolerance	High	Medium
Tunability		Yes, high	Yes, medium
Peak-to-Notch Contrast Ratio (PNCR)	Magnitude	High (>20 dB)	
	Operating Condition	Demanding (critical coupling)	Normal (branching couplers splitting ratio)
	Adjustment	Complicated (electrically driving MEMS to control gap between bus and ring)	Simple (varying power splitting ratio)
	Tolerance	Tight (matching of field transmission coefficient and amplitude attenuation factor within 3%)	Relaxed (Within 10% of 3-dB power splitting ratio)
Temperature Sensitivity		Yes, high negative TOC material, but still not compatibility with standard CMOS technology	Yes, athermal operation without extra energy consumption
Integration Potential	Compatibility with planar light wave technology	Yes	Yes
	Co-packaging Platform with SOA	Yes (hybrid)	Yes (monolithic)
Device	Size and footprint	Ultra-compact	Compact or bulk
	Fabrication, (materials, techniques, processes)	Well-developed	Established
	CMOS Compatible	Yes	Yes
	Cost	Affordable (single channel amplification in SOA), Shared (multiple channel amplification in SOA)	
	Commercial availability	Yes, increasing	Yes, widespread

7. Conclusions

In conclusion, we have presented a theoretical analysis and benchmarking of two basic optical notch filters employed to compensate for the pattern effect that manifests in an SOA due to the irregular perturbation of the SOA gain dynamics by a directly-amplified data-modulated optical signal. After following a rational procedure based on the satisfaction of signal amplitude equalization conditions, we extracted values of the critical operating parameters, which enable a fair comparison of these filters through evaluating them against appropriately-adopted figures-of-merit. In this manner, we have specified the points where the considered filters perform better due to their different principles of operation and associated spectral responses. The obtained results denote that, with the exception of larger amplitude modulation, which however is well compensated by the other comparative realized improvements, the MRR outperforms the ODI in most employed metrics. This conclusion is further supported by the comparison made between the filters from a functional and technical side. In fact, the MRR exhibits several attractive key characteristics, which, along with its capability to confront the SOA pattern effect, allows it to favorably compete against the well-established ODI. This fact constructively broadens the technological options that are available for addressing the target problem. Furthermore, it indicates that, since there is no absolute criterion for preferring one filter to the other, making a selection should be done in such a way so that it best suits the particular needs of a light wave link or network where SOA direct amplification with a suppressed pattern effect is essential. To this end, each filter should judiciously be exploited while making the necessary trade-offs according to its inherent performance advantages and disadvantages.

Author Contributions: Zoe V. Rizou conducted the simulation and interpreted the obtained results; Kyriakos E. Zoiros conceived the initial idea and wrote the paper; Antonios Hatziefremidis reviewed the results and the manuscript.

Conflicts of Interest: The authors declare no conflict of interest.

References

1. Zimmerman, D.R.; Spiekman, L.H. Amplifiers for the masses: EDFA, EDWA, and SOA amplets for metro and access applications. *J. Lightwave Technol.* **2004**, *22*, 63–70.
2. Bonk, R.; Schmuck, H.; Poehlmann, W.; Pfeiffer, T. Beneficial OLT transmitter and receiver concepts for NG-PON2 using semiconductor optical amplifiers. *J. Opt. Commun. Netw.* **2015**, *7*, A467–A473.
3. Connelly, M.J. *Semiconductor Optical Amplifiers*; Kluwer Academic Publishers: Dordrecht, The Netherlands, 2002; Chapter 6.
4. Mørk, J.; Nielsen, M.L.; Berg, T.W. The dynamics of semiconductor optical amplifiers: Modeling and applications. *Opt. Photonics News* **2003**, *14*, 42–48.
5. Kamelian. Applications Note No. 0001 "Semiconductor Optical Amplifiers (SOAs) as Power Boosters" and No. 0002 "Semiconductor Optical Amplifiers (SOAs) as Pre-Amplifiers". Available online: http://www.kamelian.com (accessed on 12 July 2017).
6. Spiekman, L.H.; Wiesenfeld, J.; Gnauck, A.; Garrett, L.; Van Den Hoven, G.; Van Dongen, T.; Sander-Jochem, M.; Binsma, J. Transmission of 8 DWDM channels at 20 Gb/s over 160 Km of standard fiber using a cascade of semiconductor optical amplifiers. *IEEE Photonics Technol. Lett.* **2000**, *12*, 717–719.
7. Udvary, E.; Berceli, T. Semiconductor optical amplifier for detection function in radio over fiber systems. *J. Lightwave Technol.* **2008**, *26*, 2563–2570.
8. Porto, S.; Antony, C.; Ossieur, P.; Townsend, P.D. An upstream reach-extender for 10 Gb/s PON applications based on an optimized semiconductor amplifier cascade. *Opt. Express* **2012**, *20*, 186–191.
9. Yiannopoulos, K.; Sagias, N.C.; Boucouvalas, A.C. Fade mitigation based on semiconductor optical amplifiers. *J. Lightwave Technol.* **2013**, *31*, 3621–3630.
10. Miao, W.; Luo, J.; Di Lucente, S.; Dorren, H.; Calabretta, N. Novel flat datacenter network architecture based on scalable and flow-controlled optical switch system. *Opt. Express* **2014**, *22*, 2465–2472.
11. Schmuck, H.; Bonk, R.; Poehlmann, W.; Haslach, C.; Kuebart, W.; Karnick, D.; Meyer, J.; Fritzsche, D.; Weis, E.; Becker, J.; et al. Demonstration of an SOA-assisted open metro-access infrastructure for heterogeneous services. *Opt. Express* **2014**, *22*, 737–748.
12. Pato, S.V.; Meleiro, R.; Fonseca, D.; Andre, P.; Monteiro, P.; Silva, H. All-optical burst-mode power equalizer based on cascaded SOAs for 10-Gb/s EPONs. *IEEE Photonics Technol. Lett.* **2008**, *20*, 2078–2080.
13. Manning, R.; Ellis, A.; Poustie, A.; Blow, K. Semiconductor laser amplifiers for ultrafast all-optical signal processing. *J. Opt. Soc. Am. B* **1997**, *14*, 3204–3216.
14. Zoiros, K.; Chasioti, R.; Koukourlis, C.; Houbavlis, T. On the output characteristics of a semiconductor optical amplifier driven by an ultrafast optical time division multiplexing pulse train. *Optik* **2007**, *118*, 134–146.

15. Rizou, Z.V.; Zoiros, K.E.; Hatziefremidis, A.; Connelly, M.J. Design analysis and performance optimization of a Lyot filter for semiconductor optical amplifier pattern effect suppression. *J. Sel. Top. Quantum Electron.* **2013**, *19*, 1–9.

16. Bonk, R.; Huber, G.; Vallaitis, T.; Koenig, S.; Schmogrow, R.; Hillerkuss, D.; Brenot, R.; Lelarge, F.; Duan, G.H.; Sygletos, S.; et al. Linear semiconductor optical amplifiers for amplification of advanced modulation formats. *Opt. Express* **2012**, *20*, 9657–9672.

17. Bogoni, A.; Potì, L.; Ghelfi, P.; Scaffardi, M.; Porzi, C.; Ponzini, F.; Meloni, G.; Berrettini, G.; Malacarne, A.; Prati, G. OTDM-based optical communications networks at 160 Gbit/s and beyond. *Opt. Fiber Technol.* **2007**, *13*, 1–12.

18. Winzer, P.J.; Essiambre, R.J. Advanced modulation formats for high-capacity optical transport networks. *J. Lightwave Technol.* **2006**, *24*, 4711–4728.

19. Mecozzi, A.; Mørk, J. Saturation induced by picosecond pulses in semiconductor optical amplifiers. *J. Opt. Soc. Am. B* **1997**, *14*, 761–770.

20. Inoue, K. Optical filtering technique to suppress waveform distortion induced in a gain-saturated semiconductor optical amplifier. *Electron. Lett.* **1997**, *33*, 885–886.

21. Watanabe, T.; Yasaka, H.; Sakaida, N.; Koga, M. Waveform shaping of chirp-controlled signal by semiconductor optical amplifier using Mach-Zehnder frequency discriminator. *IEEE Photonics Technol. Lett.* **1998**, *10*, 1422–1424.

22. Yu, J.; Jeppesen, P. Increasing input power dynamic range of SOA by shifting the transparent wavelength of tunable optical filter. *J. Lightwave Technol.* **2001**, *19*, 1316–1325.

23. Wong, C.; Tsang, H. Reduction of bit-pattern dependent errors from a semiconductor optical amplifier using an optical delay interferometer. *Opt. Commun.* **2004**, *232*, 245–249.

24. Dong, J.; Zhang, X.; Wang, F.; Hong, W.; Huang, D. Experimental study of SOA-based NRZ-to-PRZ conversion and distortion elimination of amplified NRZ signal using spectral filtering. *Opt. Commun.* **2008**, *281*, 5618–5624.

25. Hussain, K.; Pradhan, R.; Datta, P. Patterning characteristics and its alleviation in high bit-rate amplification of bulk semiconductor optical amplifier. *Opt. Quantum Electron.* **2010**, *42*, 29–43.

26. Agrawal, G.P.; Olsson, N.A. Self-phase modulation and spectral broadening of optical pulses in semiconductor laser amplifiers. *IEEE J. Quantum Electron.* **1989**, *25*, 2297–2306.

27. Wang, J. Pattern Effect Mitigation Techniques for All-Optical Wavelength Converters Based on SOAs. Ph.D. Thesis, Karlsruhe Institute of Technology, Karlsruhe, Germany, 2008.

28. Zoiros, K.E.; O'Riordan, C.; Connelly, M.J. Semiconductor optical amplifier pattern effect suppression using a birefringent fiber loop. *IEEE Photonics Technol. Lett.* **2010**, *22*, 221–223.

29. Rizou, Z.; Zoiros, K.; Hatziefremidis, A.; Connelly, M. Performance tolerance analysis of birefringent fiber loop for semiconductor optical amplifier pattern effect suppression. *Appl. Phys. B Lasers Opt.* **2015**, *119*, 247–257.

30. Zoiros, K.E.; Siarkos, T.; Koukourlis, C.S. Theoretical analysis of pattern effect suppression in semiconductor optical amplifier utilizing optical delay interferometer. *Opt. Commun.* **2008**, *281*, 3648–3657.

31. Zoiros, K.E.; Rizou, Z.V.; Connelly, M.J. On the compensation of chirp induced from semiconductor optical amplifier on RZ data using optical delay interferometer. *Opt. Commun.* **2011**, *284*, 3539–3547.

32. Zoiros, K.E.; Janer, C.L.; Connelly, M.J. Semiconductor optical amplifier pattern effect suppression for return-to-zero data using an optical delay interferometer. *Opt. Eng.* **2010**, *49*, 1–4.

33. Rizou, Z.; Zoiros, K.; Hatziefremidis, A. Semiconductor optical amplifier pattern effect suppression with passive single microring resonator-based notch filter. *Opt. Commun.* **2014**, *329*, 206–213.

34. Katz, O.; Malka, D. Design of novel SOI 1 × 4 optical power splitter using seven horizontally slotted waveguides. *Photonics Nanostruct.* **2017**, *25*, 9–13.

35. Malka, D.; Peled, A. Power splitting of 1 × 16 in multicore photonic crystal fibers. *Appl. Surf. Sci.* **2017**, *417*, 34–39.

36. Ben Zaken, B.B.; Zanzury, T.; Malka, D. An 8-channel wavelength MMI demultiplexer in slot waveguide structures. *Materials* **2016**, *9*, 881.

37. Malka, D.; Sintov, Y.; Zalevsky, Z. Design of a 1 × 4 silicon-alumina wavelength demultiplexer based on multimode interference in slot waveguide structures. *J. Opt.* **2015**, *17*, 125702.

38. Malka, D.; Sintov, Y.; Zalevsky, Z. Fiber-laser monolithic coherent beam combiner based on multicore photonic crystal fiber. *Opt. Eng.* **2015**, *54*, 011007.

39. Wong, C.; Tsang, H. Filtering directly modulated laser diode pulses with a Mach-Zehnder optical delay interferometer. *Electron. Lett.* **2004**, *40*, 938–940.

40. Rabus, D.G. *Integrated Ring Resonators: The Compendium*; Spring: Berlin, Germany, 2007.

41. Sumetsky, M. Optimization of resonant optical sensors. *Opt. Express* **2007**, *15*, 17449–17457.

42. Ruege, A.C. Electro-Optic Ring Resonators in Integrated Optics for Miniature Electric Field Sensors. Ph.D. Thesis, The Ohio State University, Columbus, OH, USA, 2001.

43. Nielsen, M.L.; Mørk, J. Bandwidth enhancement of SOA-based switches using optical filtering: Theory and experimental verification. *Opt. Express* **2006**, *14*, 1260–1265.

44. Udvary, E. Off-set filtering effect in SOA based optical access network. *Radioengineering* **2016**, *25*, 26–33.

45. Yu, H.Y.; Mahgerefteh, D.; Cho, P.S.; Goldhar, J. Optimization of the frequency response of a semiconductor optical amplifier wavelength converter using a fiber Bragg grating. *J. Lightwave Technol.* **1999**, *17*, 308–315.

46. Nielsen, M.L.; Mørk, J.; Suzuki, R.; Sakaguchi, J.; Ueno, Y. Theoretical and experimental study of fundamental differences in the noise suppression of high-speed SOA-based all-optical switches. *Opt. Express* **2005**, *13*, 5080–5086.

47. Nielsen, M.L. Experimental and Theoretical Investigation of Semiconductor Optical Amplifier (SOA) Based All-Optical Switches. Ph.D. Thesis, Technical University of Denmark, Kongens Lyngby, Denmark, 2004.

48. Rizou, Z.V.; Zoiros, K.E. Performance analysis and improvement of semiconductor optical amplifier direct modulation with assistance of microring resonator notch filter. *Opt. Quantum Electron.* **2017**, *3*, 1–21.

49. Vardakas, J.S.; Zoiros, K.E. Performance investigation of all-optical clock recovery circuit based on Fabry-Pérot filter and semiconductor optical amplifier assisted Sagnac switch. *Opt. Eng.* **2007**, *46*, 1–21.

50. Gutiérrez-Castrejón, R.; Filios, A. Pattern-effect reduction using a cross-gain modulated holding beam in semiconductor optical in-line amplifier. *J. Lightwave Technol.* **2006**, *24*, 4912–4917.

51. Yan, S.; Zhang, Y.; Dong, J.; Zheng, A.; Liao, S.; Zhou, H.; Wu, Z.; Xia, J.; Zhang, X. Operation bandwidth optimization of photonic differentiators. *Opt. Express* **2015**, *23*, 18925–18936.

52. Agrawal, G.P. *Fiber-Optic Communication Systems*, 3rd ed.; Wiley: New York, NY, USA, 2002.

53. Bonk, R.; Vallaitis, T.; Guetlein, J.; Meuer, C.; Schmeckebier, H.; Bimberg, D.; Koos, C.; Freude, W.; Leuthold, J. The input power dynamic range of a semiconductor optical amplifier and its relevance for access network applications. *IEEE Photonics J.* **2011**, *3*, 1039–1053.

54. Vacondio, F.; Ghazisaeidi, A.; Bononi, A.; Rusch, L.A. Low-complexity compensation of SOA nonlinearity for single-channel PSK and OOK. *J. Lightwave Technol.* **2010**, *28*, 277–288.

55. Xu, J.; Zhang, X.; Mørk, J. Investigation of patterning effects in ultrafast SOA-based optical switches. *IEEE J. Quantum Electron.* **2010**, *46*, 87–94.

56. MacWilliams, F.J.; Sloane, N.J. Pseudo-random sequences and arrays. *Proc. IEEE* **1976**, *64*, 1715–1729.

57. Singh, S.; Kaler, R. Transmission performance of 20 × 10 Gb/s WDM signals using cascaded optimized SOAs with OOK and DPSK modulation formats. *Opt. Commun.* **2006**, *266*, 100–110.

58. Bhardwaj, A.; Doerr, C.R.; Chandrasekhar, S.; Stulz, L.W. Reduction of nonlinear distortion from a semiconductor optical amplifier using an optical equalizer. *IEEE Photonics Technol. Lett.* **2004**, *16*, 921–923.

59. Niehusmann, J.; Vörckel, A.; Bolivar, P.H.; Wahlbrink, T.; Henschel, W.; Kurz, H. Ultrahigh-quality-factor silicon-on-insulator microring resonator. *Opt. Lett.* **2004**, *29*, 2861–2863.

60. Fujiwara, M.; Imai, T.; Taguchi, K.; Suzuki, K.I.; Ishii, H.; Yoshimoto, N. Field trial of 100-km reach symmetric-rate 10 G-EPON system using automatic level controlled burst-mode SOAs. *J. Lightwave Technol.* **2013**, *31*, 634–640.

61. Porzi, C.; Kado, Y.; Shimizu, S.; Maruta, A.; Wada, N.; Bogoni, A.; Kitayama, K.I. Simple uplink SOA pattern effects compensation for reach-extended 10 G-EPONs. *IEEE Photonics. Technol. Lett.* **2014**, *26*, 165–168.

62. Kim, H. 10-Gb/s operation of RSOA using a delay interferometer. *IEEE Photonics Technol. Lett.* **2010**, *22*, 1379–1381.

63. Elrefaie, A.F.; Wagner, R.E.; Atlas, D.; Daut, D. Chromatic dispersion limitations in coherent light wave transmission systems. *J. Lightwave Technol.* **1988**, *6*, 704–709.

64. Forghieri, F.; Prucnal, P.; Tkach, R.; Chraplyvy, A.R. RZ versus NRZ in nonlinear WDM systems. *IEEE Photonics Technol. Lett.* **1997**, *9*, 1035–1037.

65. Levaufre, G.; Le Liepvre, A.; Jany, C.; Accard, A.; Kaspar, P.; Brenot, R.; Make, D.; Lelarge, F.; Olivier, S.; Malhouitre, S.; et al. Hybrid III-V/silicon tunable laser directly modulated at 10 Gbit/s for short reach/access networks. In Proceedings of the European Conference on Optical Communication, Cannes, France, 21–25 September 2014.

66. Dubé-Demers, R.; LaRochelle, S.; Shi, W. Ultrafast pulse-amplitude modulation with a femtojoule silicon photonic modulator. *Optica* **2016**, *3*, 622–627.

67. Gutiérrez-Castrejón, R.; Occhi, L.; Schares, L.; Guekos, G. Recovery dynamics of cross-modulated beam phase in semiconductor amplifiers and applications to all-optical signal processing. *Opt. Commun.* **2001**, *195*, 167–177.

68. Gayen, D.K.; Chattopadhyay, T.; Zoiros, K.E. All-optical D flip-flop using single quantum-dot semiconductor optical amplifier assisted Mach-Zehnder interferometer. *J. Comput. Electron.* **2015**, *14*, 129–138.

69. Zhang, L.; Li, Y.; Song, M.; Beausoleil, R.G.; Willner, A.E. Data quality dependencies in microring-based DPSK transmitter and receiver. *Opt. Express* **2008**, *16*, 5739–5745.

70. Tazawa, H.; Kuo, Y.H.; Dunayevskiy, I.; Luo, J.; Jen, A.K.Y.; Fetterman, H.R.; Steier, W.H. Ring resonator-based electrooptic polymer traveling-wave modulator. *J. Lightwave Technol.* **2006**, *24*, 3514–3519.

71. Ye, T.; Cai, X. On power consumption of silicon-microring-based optical modulators. *J. Lightwave Technol.* **2010**, *28*, 1615–1623.

72. Madsen, C.K.; Zhao, J.H. *Optical Filter Design and Analysis: A Signal Processing Approach*; John Wiley & Sons, Inc.: New York, NY, USA, 1999; pp. 1–5.

73. Malka, D.; Cohen, M.; Turkiewicz, J.; Zalevsky, Z. Optical micro-multi-racetrack resonator filter based on SOI waveguides. *Photonics Nanostruct.* **2015**, *16*, 16–23.

74. Wu, Q.; Zhang, H.; Fu, X.; Yao, M. Spectral encoded photonic analog-to-digital converter based on cascaded unbalanced MZMs. *IEEE Photonics Technol. Lett.* **2009**, *21*, 224–226.

75. Bogaerts, W.; De Heyn, P.; Van Vaerenbergh, T.; De Vos, K.; Kumar Selvaraja, S.; Claes, T.; Dumon, P.; Bienstman, P.; Van Thourhout, D.; Baets, R. Silicon microring resonators. *Laser Photonics Rev.* **2012**, *6*, 47–73.

76. Kim, T.Y.; Hanawa, M.; Kim, S.J.; Hann, S.; Kim, Y.H.; Han, W.T.; Park, C.S. Optical delay interferometer based on phase shifted fiber Bragg grating with optically controllable phase shifter. *Opt. Express* **2006**, *14*, 4250–4255.

77. Li, J.; Worms, K.; Maestle, R.; Hillerkuss, D.; Freude, W.; Leuthold, J. Free-space optical delay interferometer with tunable delay and phase. *Opt. Express* **2011**, *19*, 11654–11666.

78. Su, T.; Zhang, M.; Chen, X.; Zhang, Z.; Liu, M.; Liu, L.; Huang, S. Improved 10-Gbps uplink transmission in WDM-PON with RSOA-based colorless ONUs and MZI-based equalizers. *Opt. Laser Technol.* **2013**, *51*, 90–97.

79. Luo, Z.C.; Cao, W.J.; Luo, A.P.; Xu, W.C. Polarization-independent, multifunctional all-fiber comb filter using variable ratio coupler-based Mach-Zehnder interferometer. *J. Lightwave Technol.* **2012**, *30*, 1857–1862.

80. Bhardwaj, A.; Sauer, N.; Buhl, L.; Yang, W.; Zhang, L.; Neilson, D.T. An InP-based optical equalizer monolithically integrated with a semiconductor optical amplifier. *IEEE Photonics Technol. Lett.* **2007**, *19*, 1514–1516.

81. Absil, P.; Hryniewicz, J.; Little, B.; Wilson, R.; Joneckis, L.; Ho, P.T. Compact microring notch filters. *IEEE Photonics Technol. Lett.* **2000**, *12*, 398–400.

82. Lee, M.C.; Wu, M.C. MEMS-actuated microdisk resonators with variable power coupling ratios. *IEEE Photonics Technol. Lett.* **2005**, *17*, 1034–1036.

83. Zhou, Z.; Yin, B.; Deng, Q.; Li, X.; Cui, J. Lowering the energy consumption in silicon photonic devices and systems. *Photonics Res.* **2015**, *3*, B28–B46.

84. Kaspar, P.; Brenot, R.; Le Liepvre, A.; Accard, A.; Make, D.; Levaufre, G.; Girard, N.; Lelarge, F.; Duan, G.H.; Pavarelli, N.; et al. Packaged hybrid III-V/silicon SOA. In Proceedings of the European Conference on Optical Communication, Cannes, France, 21–25 September 2014.

85. Li, Y.; Zhang, Y.; Zhang, L.; Poon, A.W. Silicon and hybrid silicon photonic devices for intra-datacenter applications: State of the art and perspectives. *Photonics Res.* **2015**, *3*, B10–B27.

86. Gunn, C. CMOS photonics for high-speed interconnects. *IEEE Micro* **2006**, *26*, 58–66.

87. Tan, H.N.; Matsuura, M.; Kishi, N. Enhancement of input power dynamic range for multiwavelength amplification and optical signal processing in a semiconductor optical amplifier using holding beam effect. *J. Lightwave Technol.* **2010**, *28*, 2593–2602.

88. Padmaraju, K.; Bergman, K. Resolving the thermal challenges for silicon microring resonator devices. *Nanophotonics* **2014**, *3*, 269–281.

89. Ding, Y.; Pu, M.; Liu, L.; Xu, J.; Peucheret, C.; Zhang, X.; Huang, D.; Ou, H. Bandwidth and wavelength-tunable optical bandpass filter based on silicon microring-MZI structure. *Opt. Express* **2011**, *19*, 6462–6470.

applied
sciences

MDPI

Article

Estimation of the Performance Improvement of Pre-Amplified PAM4 Systems When Using Multi-Section Semiconductor Optical Amplifiers

Seán P. Ó Dúill *, Pascal Landais and Liam P. Barry

Radio and optics research laboratory, School of Electronic Engineering,
Dublin City University, Dublin 9 D09 W6Y4, Ireland; pascal.landais@dcu.ie (P.L.); liam.barry@dcu.ie (L.P.B.)
* Correspondence: sean.oduill@dcu.ie; Tel.: +353-1-700-5869

Received: 1 August 2017; Accepted: 31 August 2017; Published: 5 September 2017

Featured Application: Multi-section semiconductor optical amplifiers are known to have superior noise and gain saturation performances compared to regular single section semiconductor optical amplifiers. In this paper, we estimate the performance benefit of using multi-section semiconductor optical amplifiers as pre-amplifiers for short range optical communication links within datacenters.

Abstract: Multi-section semiconductor optical amplifiers (SOA) have been shown to have superior noise and linearity performance compared with single section SOAs. We show how to create a simplified numerical model for multi-section SOAs that is suitable for optical communication system simulations and use that model to investigate the amplification performance of 56 Gbit/s four-level pulse amplitude modulation signals. We find that a multi-section SOA could provide an improvement in input power dynamic range exceeding 3 dB compared to a single section SOA that has the same unsaturated gain.

Keywords: semiconductor optical amplifier; pulse amplitude modulation; noise

---

1. Introduction

There have been tremendous advances in the information carrying capacity of optical communication systems within the past decade, with the optical techniques for communication now migrating to intra-datacenter communication [1,2]. The growth of stored data within datacenters and the necessity to retrieve this data is driving the requirement for cost-effective, higher capacity intra-datacenter interconnects to 400 Gbit/s and the data modulation format is evolving towards four-level pulse amplitude modulation (PAM4) that carries two bits per every transmitted symbol [2]. Figure 1a presents a simplified schematic of a data connection between two clusters of racks in a datacenter where optoelectronic transponders and optical fiber form the basic communication link. Already, a bank of eight externally modulated lasers using electro-absorption modulators to deliver 400 Gbit/s links on a single fiber can be offered with 28 Gbaud PAM4 offering a raw data rate of 56 Gbit/s per wavelength [3].

Semiconductor optical amplifiers (SOA) are low cost, highly integratable amplifiers that may find two roles within intra-datacenter communication. Firstly, optical switches with light paths defined by activated SOAs have been demonstrated [4,5] and could provide additional switching functionality in an optical-domain switching layer. Secondly, SOAs are being given serious consideration to also play a role as pre-amplifiers for PAM4 signals as reach extenders or to overcome inevitable insertion losses of an optical switch as shown in Figure 1b [6,7]. Usually, SOAs suffer from a relatively high noise figure (NF) which limits the minimum signal power that can be reliably amplified to a few tens of μW;

SOAs also suffer from low gain saturation power (the order of 10 mW) which distorts the amplified signal even when the average input signal power is of the order of hundreds of microwatts [8,9].

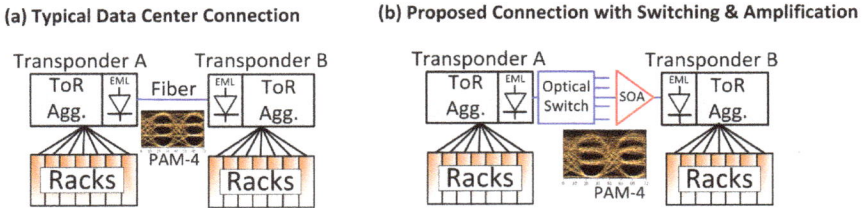

(a) Typical Data Center Connection   **(b) Proposed Connection with Switching & Amplification**

Figure 1. (**a**) Simplified schematic showing communication between two clusters of racks in a datacenter. The transponders use externally modulated lasers (EML) to communicate at up to 400 Gbit/s between the top of the rack aggregators (ToR Agg.); (**b**) Proposed use of semiconductor optical amplifiers (SOA) as a pre-amplifier to overcome losses within an optical switch. Eyediagrams of the simulated PAM4 signals are shown, in (**b**) a slightly distorted eyediagram due to the SOA nonlinearities are shown.

The NF and saturation power of SOAs can be improved by subdividing the amplifier into separate sections and applying a different bias current to each subsection [10]. The concept of a multi-section SOA [10] is shown in Figure 2. In Figure 2a, for a single-section SOA, a large DC bias current is applied to the SOA resulting in equal carrier injection being applied along the entire length of the SOA. For the multi-contact SOA, the SOA is subdivided into smaller sections with each section biased independently. The SOA is operated with favorable current biasing arrangements, whereby a stronger bias current is applied towards the output facet allowing for an improvement of the effective NF and increase in the effective saturation power relative to single-section SOAs [10–12]. This favorable biasing current arrangement can be done without increasing the overall injected current and hence keeping the power consumption constant. In separate works, regarding reflective SOAs (RSOA), with the RSOA subdivided into two subsections was found to have improved the direct modulation electro-optic response of the RSOA [13]. To date, no study (neither experimental nor numerical) has been performed on investigating the benefits of amplifying information-carrying modulated optical signals using multi-section SOAs, only the NF and saturation powers have been measured and/or calculated [10–12]. In addition, a simplified numerical model for multi-section SOAs has not yet been developed that would allow estimating the amplification performance of modulated signals in multi-section SOAs.

In this paper, we show how to create a simplified multi-section SOA model and use that model to evaluate the amplification performance of multi-section SOAs, and as comparison to conventional single-section SOAs for 28 Gbaud PAM4 signals. We are modifying our simplified SOA model in [14] to consider multi-section SOAs. We find that in order to replicate the experimental results for amplification of PAM4 signals using SOAs in [15], we need only consider amplified spontaneous emission (ASE) noise generated in the SOA and thermal noise in an optical receiver, and we achieve similar bit-error rate (BER) performance to experimental 28 Gbaud PAM4 systems [15]. The input power dynamic range (IPDR) is an important metric for amplified systems using SOAs [8], because the IPDR indicates the useful input power range over which signals can be amplified and the data retrieved without error [8]. We show that a multi-section SOA could improve the IPDR by more than 3 dB relative to the IPDR of a similar gain single-section SOA. The results, from this simple system simulator for multi-section SOAs, motivate the further experimental investigation of multi-section SOAs for datacom applications.

(a) Bias arrangement for single-section SOA

Bias Current

Light IN

Light OUT

SOA

(b) Bias arrangement for multi-section SOA

Section Bias Current

Light IN

Current Blocker

Current Blocker

Light OUT

SOA

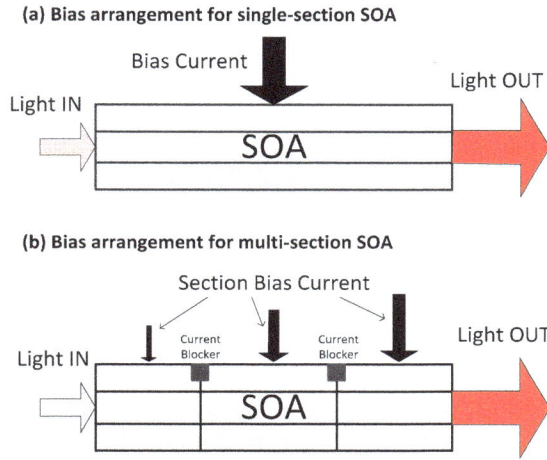

Figure 2. (**a**) Simplified schematic of a travelling wave single-section SOA where the population inversion is created by biasing the whole amplifier with a single current; (**b**) Multi-section SOA, the bias currents to the smaller amplifier sections are independently set. The multi-section biasing arrangement creates an SOA with improved noise and gain saturation performances compared to the case in (**a**).

2. Numerical Simulator

We create an optical field, E, that is normalised such that $|E|^2$ represents the optical power. The four level 28 Gbaud PAM4 signal is encoded within the optical power domain. We set the extinction ratio between the strongest and weakest levels of the PAM4 signal to be 8 dB. Before describing the SOA model we first describe the linear photo-detector model as depicted in Figure 3a. The received photo-current is given as

$$i_{RX} = R|E_{rec}(t)|^2 \tag{1}$$

where E_{rec} is the optical field incident on the photodiode, the detector responsivity $R = 1$ A/W throughout this paper. In the detector, electrical thermal noise is added to the current generated by the signal, the thermal noise n_{RX} is considered to be white Gaussian noise with autocorrelation [16]

$$\int_{-\infty}^{\infty} n_{RX}(\tau)n_{RX}(\tau + t)d\tau = \frac{4kT}{R_{RX}}\delta(t). \tag{2}$$

With k being Boltzmann's constant, T (K) the temperature, and $R_{RX} = 50\ \Omega$ is the impedance of the receiver, and δ is the Dirac delta function. The combined signal and noise is low-pass filtered and then a decision is made on the transmitted PAM4 symbol. Error counting after comparing the received and transmitted bits determines the BER. In the simulator the current waveform is generated by generating a random waveform with values taken from a zero-mean, unity variance Gaussian distribution,

$$\overline{n}_{RX} = \sqrt{\frac{4kT}{R_{RX}}B_{sim}}e_{elec}. \tag{3}$$

e_{elec} is a random noise waveform whose values are taken from a zero-mean, unity-variance Gaussian distribution, and B_{sim} is the simulation bandwidth (i.e., inverse sampling time) and equals 1 THz throughout this paper. A different value for e_{elec} is generated for each sampling point in the waveform for the PAM4 signal E.

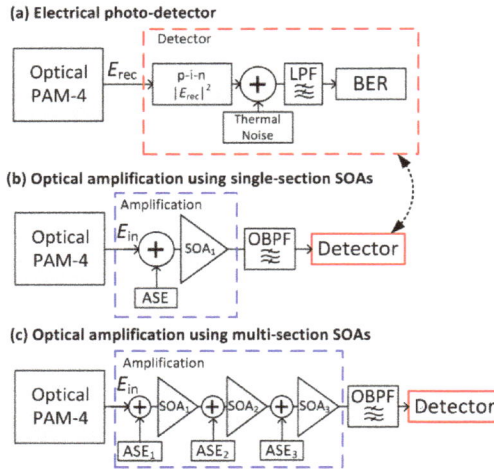

Figure 3. Simulation scenarios performed in this submission. (**a**) Linear receiver with PIN photodiode including thermal noise. LPF: Lowpass filter; (**b**) Pre-amplified PAM4 signal using an SOA. ASE is added before SOA, the receiver is as in (**a**); (**c**) multi-section SOA, the SOA is split into three separate SOA subsections, ASE noise is added before each SOA subsection. Adjusting the bias can alter the NF and saturation power of each subsection. The detector is as in (**a**).

The amount of thermal noise remains fixed and the BER changes by changing the received signal power $|E_{rec}|^2$. We calculate the BER for different values of received optical signal power with the results shown in Figure 4. We indicate the sensitivity to achieve a BER of 3.16×10^{-4} because this is the BER at which the forward error correction algorithms can effectively yield error-free performance [2]. The received power that attains a BER of 3.16×10^{-4} is estimated to be -11.4 dBm, which is in line with an experimentally obtained sensitivity of -10.8 dBm [15], and within an acceptable range for receiver sensitivity. Reducing the responsivity of the photodiode to 0.7 would add +1.5 dB to our calculated receiver sensitivity. The detector model is central to all of our simulations and we use the detector model after amplifying the signal in an SOA.

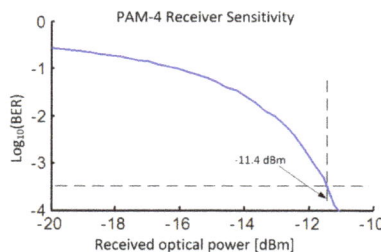

Figure 4. Receiver bit-error rate (BER) performance for 28 Gbaud PAM4. The ER is 7.6 dB. The sensitivity at BER 3×10^{-4} occurs at -11.8 dBm.

The simulation models for the SOAs are shown in Figure 3b,c. Typically, SOAs are travelling wave models and consideration for the spatially-distributed gain should be considered, though in the absence of internal scattering losses the SOA can be equivalently treated as a lumped device [17] without needing to consider the propagation of the optical field in the SOA. The lumped SOA model approach avoids the need to subdivide the SOA into smaller subsections in order to compute the propagation of the optical field. Additionally, the lumped SOA model approach requires that only a single differential equation needs to be numerically solved to estimate the SOA gain. The computational savings by adopting the lumped SOA model approach makes that approach attractive for use within communication system simulators, where tens of thousands of data symbols need to be simulated in order to gauge meaningful system performance. An example of using the lumped SOA model approach for communication system simulators is given in [18]. To address the issue of including non-negligible internal scattering losses we developed an improved lumped SOA model [14] that is accurate to ~1 dB and we have used this model to estimate the wavelength conversion performance of optical signals in SOAs [19]. The SOA dynamics are described by a single differential equation for each subsection [14]:

$$\frac{dh_i}{dt} = \frac{h_{0,i} - h_i}{\tau_S} - \frac{h_i}{h_i - \alpha_{loss,i}} \frac{\exp(h_i - \alpha_{loss,i}) - 1}{P_{sat}\tau_S} |E_{in}(t)|^2. \tag{4}$$

With h representing the integrated gain-coefficient over the length of the SOA section, h_0 is the unsaturated value of the integrated gain coefficient, τ_S is the carrier lifetime, α_{loss} is the integrated scattering losses, P_{sat} is the saturation optical power at which the gain coefficient reaches half of the unsaturated value, and $|E_{in}(t)|^2$ is the incident optical power to the SOA. The second term on the right hand side of (4) describes stimulated emission with the first term on the right hand side describing gain recovery due a replenishment of the gain via the bias current. The output field from the SOA subsection is

$$E_{out,i}(t) = E_{in,i}(t) \exp\left(\frac{1}{2}\{(1 - j\alpha_H)h_i(t) - \alpha_{loss,i}\}\right). \tag{5}$$

α_H being the gain-phase coupling parameter. The 'i' index in (4) and (5) refer to a subsection in the multi-section SOA. To account for the ASE generated within the SOA we add an equivalent ASE noise to the optical field before passing the field through the SOA [18]. The equivalent optical ASE noise n_{ASE} is assumed to be white Gaussian noise with the following autocorrelation:

$$\int_{-\infty}^{\infty} n_{ASE}(\tau)n_{ASE}^*(t+\tau)d\tau = n_{sp}h\upsilon \frac{\exp(h_{0,i} - \alpha_{loss,i}) - 1}{\exp(h_{0,i} - \alpha_{loss,i})}\delta(t). \tag{6}$$

With $h\upsilon$ being the photon energy, n_{sp} is the gain inversion parameter and is related to the NF by NF $= \log_{10}(2n_{sp})$. There is a correction factor to ensure that the ASE noise power equals $n_{sp}h\upsilon(\exp[h_{0,i} - \alpha_{loss,i}] - 1)$ at the output [18]. This correction factor is accurate while the SOA gain remains unsaturated, though the ASE noise only affects the performance of low-power input signals up to a few tens of μW. In the numerical simulator we generate two random waveforms with samples taken from two uncorrelated, zero-mean, unity-variance Gaussian distributions $e_{ASE-I,i}$ and $e_{ASE-Q,i}$ to represent the in-phase (I) and quadrature (Q) components of the optical field

$$\overline{n}_{ASE,i} = \sqrt{\frac{n_{sp}h\upsilon}{2}\frac{\exp(h_{0,i}) - 1}{\exp(h_{0,i})}} B_{sim}\left(e_{ASE-I,i}(t) + je_{ASE-Q,i}(t)\right). \tag{7}$$

For the multi-section SOA, the ASE generated within each section is uncorrelated.

We consider the SOA as a pre-amplifier, and typically SOA pre-amplifiers have moderate gains of 13 to 15 dB amplification [8–11] in order to avoid deleterious gain saturation. Selecting a device with a larger gain will only reduce the input optical power at which gain saturation occurs. The lower power limit of amplifying signals in SOAs is determined by the NF (or effective NF for the case of multi-section SOAs) and mainly independent of the SOA gain. For specific parameter selection, we aim to replicate SOA saturation characteristics of an actual multi-section SOA [10]. We set both (multi-section and single section) SOAs to have a total unsaturated gain of 13 dB. For the single section SOA: $h_0 = 6$, $\alpha_{loss} = 3$, this implies that the unsaturated power gain is

$$\exp[h_0 - \alpha_{loss}] = 20. \tag{8}$$

The NF is large at 8 dB, $\tau_S = 200$ ps, $P_{sat} = 20$ mW. Note, that for the single section SOA, only a single differential equation of the type in (4) is solved. The breakdown of the parameters for each section in the multi-section SOA is given in Table 1. We assume that that the biasing arrangement allows for the first section to have modest gain with a small NF of 4 dB, and the third section to have the largest gain with a large NF of 9 dB. The multi-section SOA has an overall net gain of

$$\exp\left[\sum_i (h_{0,i} - \alpha_{loss,i})\right] = 20. \tag{9}$$

Table 1. List of SOA parameters.

	Multi Section SOA, Section 1	Multi Section SOA, Section 2	Multi Section SOA, Section 3	Single Section SOA
h_0	1.5	2	2.5	6
α_{loss}	1	1	1	3
α_H	3	3	3	3
P_{sat}	15 mW	20 mW	45 mW	20 mW
τ_S	260 ps	200 ps	88 ps	200 ps
NF	4 dB	8 dB	9 dB	8 dB

The saturation power P_{sat} is not a unique quantity to the SOA though rather the saturation energy which is the product $P_{sat}\tau_S$ remains constant [17]. Under strong bias currents, the carrier density increases and the carrier lifetime τ_S falls due to increased bimolecular and Auger recombination, therefore we adjust the saturation powers such that the product $P_{sat}\tau_S = 4$ pJ for the single-section SOA and for all subsections of the multi-section SOA.

The SOA gain saturation provides much insight into the improvements offered by the multi-section SOA. We will calculate the SOA gain–output power relations for both SOAs. We create continuous wave fields without any modulation and calculate the steady-state optical power gain. Note that stochastic effects are avoided by setting the NF = −50 dB in order to get clear and precise gain calculations. The results are shown in Figure 5. For low output powers below −5 dBm the unsaturated gain for both SOAs is 13 dB. The SOA gain falls as the output increases and we note that P_{3dB} is the output power at which the gain has fallen from the unsaturated value by 3 dB. The single section SOA has a P_{3dB} of 8.28 dBm, whereas the P_{3dB} for the multi-section SOA is 10.74 dBm. We also plot on Figure 5 the gain saturation performance when the biasing arrangements are swapped between the first and third sections, the output saturation power reduces to 7.1 dBm. Such a biasing arrangement might be favourable for enhancing the nonlinear behaviour of the SOA and could be exploited for all-optical signal processing purposes [19], though amplification using this configuration is not explored here. Experimental results taken from a multi-section SOA show a P_{3dB} of 9.3 dBm for the high saturation power configuration and a P_{3dB} of 6.3 dBm for the swapped biasing arrangement [10]. We next examine how the SOAs perform when amplifying PAM4 signals.

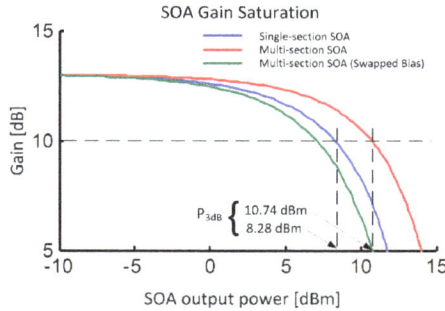

Figure 5. SOA gain saturation curves for the regular single-section SOA (blue) and multi-section SOA in red. The multi-section SOA is biased so that the P_{3dB} is greater and increases from 8.28 dBm for the single-section SOA to 10.74 dBm for the multi-contact SOA. We find that the results show similar performance to single section and multi section SOAs in [10,11]. The green curve shows the gain saturation characteristic when the bias to the first and third sections are swapped; in this scenario the effective gain saturation is much stronger resulting in a P_{3dB} of 7.1 dB.

3. Single Channel PAM4 Amplification Results

We create an optical field with 20,000 PAM4 symbols (40,000 bits) at 28 Gbaud encoded into the power waveform. The signals pass through the SOA and are detected in the receiver. The optical input power to the SOA is varied from −30 dBm to 0 dBm, the output of the SOA passes through an optical bandpass filter (OBPF) before the detector block which calculates the BER. The OBPF has a super-Gaussian spectral profile with optical field transfer function

$$H_{OBPF}(f) = \exp\left(-\frac{1}{4}\left(\frac{f-f_0}{\Delta f_{OBPF}}\right)^2\right). \tag{10}$$

With f_0 being the central frequency of the OBPF, and Δf_{OBPF} = 50 GHz, resulting in a full-width half maximum bandwidth of 108 GHz. This is wide enough to pass the central channel while offering strong rejection of neighboring channels in a multi-channel simulation, more details about the OBPF are given in the following section. Note that the smallest frequency difference to the central channel of interest that we consider throughout this paper is 100 GHz. The OBPF also limits the amount of ASE that reaches the photo-detector and reduces the detrimental spontaneous-spontaneous beat noise in direct detection systems [16]. The BER results are shown in Figure 6 for a single channel scenario. The BER is large for low input powers below −30 dBm because the ASE dominates, and the BER reduces sharply as the power is increased. The BER is just greater than 10^{-3} at −20 dBm which is in-line with experimental results [15]. The multi-section SOA gives a 0.84 dB sensitivity improvement on the lower power side compared to a single-section SOA due to the lower effective NF of the multi-section SOA. As the power increases, the ASE no longer limits performance and BERs $<10^{-4}$ are achieved. However, as the power increases above −5 dBm, the BER increases sharply due to signal distortion caused by the SOA gain saturation. The multi-section SOA performs better than a single section SOA because of increased effective P_{sat}. These findings are in line with an experimental study of the saturation characteristics of multisection SOAs that showed picosecond pulses suffer less distortion than regular single section SOAs [20]. We use the IPDR to explicitly quantify the improvement in the signal quality after SOA amplification. The IPDR gives the range of input powers over which the BER remains below that of the forward error correcting algorithm. The calculated IPDR is 13.55 dB for single section SOA and 16.85 dB for a multi-section SOA resulting in a 3.3 dB improvement in the IPDR. To visualize the signal quality at the output of the SOA we plot the eyediagrams of the PAM4 signal for different input powers to the multi-section SOA. In Figure 7a, we show the eyediagram for a low input power of −21 dBm, the signal is corrupted by ASE including signal-spontaneous beat noise which is clearly worse for the stronger levels of the PAM4 signal. In Figure 7b, we increase the SOA input power to −11 dBm and we find a clean PAM4 signal at the SOA output and no bit errors were detected for this input power. In Figure 7c, as the power is increasing to −6 dBm, we see that signal distortion begins to occur and when the power is increased further to 0 dBm we find that the SOA gain saturation corrupts the PAM4 signal resulting in increased BER. The signal distortion at higher input power levels is due to deterministic SOA gain saturation and signal pre-distortion techniques can be applied when modulating the PAM4 signal at the transmitter in order to avoid signal distortion [7]. The biggest source of detection errors occurs for the upper levels, as evident in Figure 7d; and therefore increasing the separation of the upper levels of the PAM4 signal would improve the BER performance at the higher input power levels.

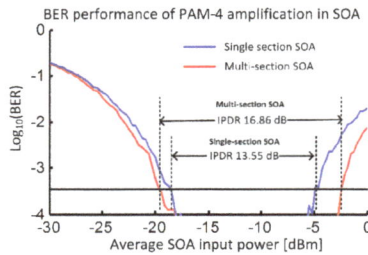

Figure 6. BER performance of pre-amplified 28 Gbaud PAM4 signals using a single section SOA (blue) and multi-section SOA (red). The multi-section SOA outperforms the single SOA with improved IPDR from 12.91 dB to 16.04 dB.

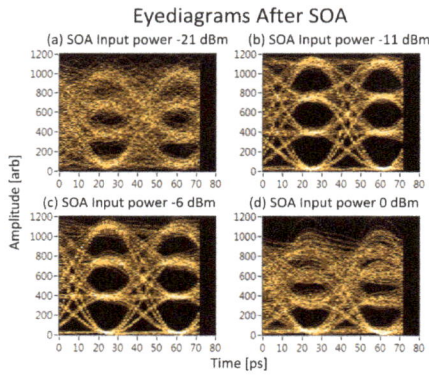

Figure 7. We display eyediagrams derived from the output intensity waveforms from the SOA. The eyediagrams are taken over a wide range of input power levels to the SOA in order to display different signal impairments. (**a**) −21 dBm; in this power regime the ASE noise dominates, note the larger splitting of the upper stronger PAM4 symbols due to signal-spontaneous beat noise; (**b**) −11 dBm; the eyediagram is clearly open with four distinct levels and no errors were detected in the 20,000 symbols used in the simulation. In this power regime, the signal is strong enough to overcome the ASE noise added by the SOA while being weak enough to not induce nonlinear gain saturation; (**c**) −6 dBm; in this power regime, there is little influence of the ASE though the signal is just strong enough to begin inducing distortions by saturating the gain of the amplifier; (**d**) 0 dBm; in this power regime, the clearly visible signal distortions are due to strong gain saturation within the SOA.

4. Multichannel PAM4 Amplification Results

Data needs to be multiplexed upon multiple wavelengths to provide for high bit rate connections of 200 Gbit/s and 400 Gbit/s [2,3,15], and eight separate wavelengths with 28 Gbaud PAM4 signals modulated on each are required to reach 400 Gbit/s [3]. Our SOA model can handle multiple channels, though there is no inherent wavelength dependence of the SOA gain, refractive index, and saturation power in the model. In the absence of wavelength dependence in the SOA model, we limit ourselves to just four channels. The additional channels with each detuned from each other by a sufficient amount to avoid spectral spillage into neighbouring channels and the frequency detunings must be small enough such that the entire multi-channel bandwidth is smaller than the simulation bandwidth of 1 THz. Though, typically for datacenter applications a coarse wavelength grid of 4.5 nm (~800 GHz) is used allowing for large frequency offsets from the center of each wavelength band [21], here we concentrate on smaller channel spacings because of the limited simulation bandwidth and lack of wavelength dependent gain and gain saturation in our model. In the complex valued optical field representation the total multi-channel field is given by:

$$E_{tot} = \sum_{0}^{K-1} E_k \exp\{j2\pi\Delta f_k t\} \tag{11}$$

where E_k is the optical field of the kth channel and Δf_k is the frequency detuning of the kth channel from the central frequency channel and K is the number of channels. We concentrate on gathering BER results for one channel (results already shown in Section 3); for two channels when the detuning frequencies are 0 and 100 GHz; and for four channels when the detuning frequencies are: 0, −100, 120, and −220 GHz. The input spectrum to the SOA is shown for four channels in Figure 8a, and each channel has an average power of 1 mW, the spectral profile of the OBPF is also shown. The filtered signal after amplification is shown in Figure 8b to show strong rejection of neighbouring channels. These channels are well within the 1 THz simulation bandwidth with unequal channel spacings used to avoid signal impairments due to interference from four-wave mixing components. The ASE generated using (7) is already wideband over the entire simulation bandwidth and no further modifications are necessary.

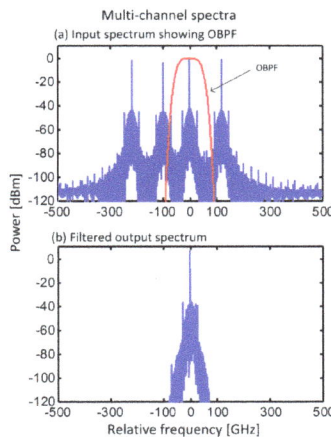

Figure 8. (**a**) SOA input spectrum showing four wavelength channels modulated with PAM4 data. The average power per channel is 1 mW. The BER estimates are performed on the channel that has been filtered out using the OBPF shown in red. (**b**) Spectrum of the filtered signal at the SOA output showing strong rejection of neighboring channels.

We generate an optical field for one, two, and four channels using (11), a different (uncorrelated) PAM4 signal is applied to each channel and the channels are temporally delayed by a random constant delay such that the symbol transition occurring for each channel are not synchronized, i.e., both waveforms with uncorrelated symbols have non-synchronized symbol transitions, and this helps to average out the optical intensity variations and reduce the signal distortion in the SOA. The SOA gain conditions are the same as in Section 3 for both single-section and multi-section SOAs.

The BER results of the multi-channel amplification are shown in Figure 9. For both SOA configurations, the BER performance is the same irrespective of the number of co-propagating channels for low input powers between −30 dBm and −18 dBm, this is unsurprising since there is no gain saturation occurring and the BER is dominated by ASE noise. When the ASE noise limits the performance, the BER of the multi-section SOA is smaller than the BER of the single-section SOA because of the lower effective NF of the multi-section SOA. Experimentally, increasing the number of co-propagating channels did not change the BER performance at low-input power levels [15], where the BER of the central channel remained the same with no other co-propagating channels and with two other co-propagating channels. In our simulations, for the case of amplifying at higher input powers with significant gain saturation the BER worsens with increasing number of channels. There is a 3 dB reduction in the IPDR for the multi-section SOA when comparing with the amplification of four channels as opposed to amplifying a single channel; and a 2 dB reduction in the IPDR for the single-section SOA when amplifying four channels as opposed to amplifying a single channel. Nonetheless, the performance of the multi-section SOA remains better due to the larger effective saturation power of the multi-section SOA. Also from Figure 9 we notice that there seems to be a cross-over point in the BER performance when the multi-channel BER outperforms the single channel BER; for the single section SOA this occurs at −4 dBm, and for the multi-section SOA this occurs above 0 dBm. The crossover point occurs because the power fluctuations are smoothed out due to the increased number of channels that results in reduced intensity fluctuations on individual channels. Though in comparison with the experimental results for amplifying PAM4 signals in SOA [15] for three channel amplification: the BER performance in [15] is worse than presented here. The channel spacings used in [15] were much greater than those considered here and the discrepancies could be due to the inevitable wavelength dependent gain and gain saturation characteristics of SOAs. Without knowledge of the exact SOA gain saturation characteristics, the channels could be closer to the gain-peak of the SOA and therefore the saturation power would be lower such that different channels will experience a different saturation power. An experimental study for amplifying a single wavelength channel and two wavelength channel systems with only 150 GHz channel separation for various modulation formats (though PAM4 as not one of them) [9] showed one or two dB power penalty when amplifying two channels compared to the single channel scenario, and the simulation results here are in qualitative agreement those multi-channel penalties.

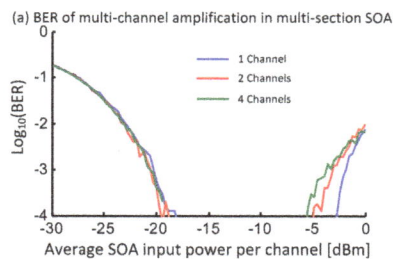

(a) BER of multi-channel amplification in multi-section SOA

Figure 9. *Cont.*

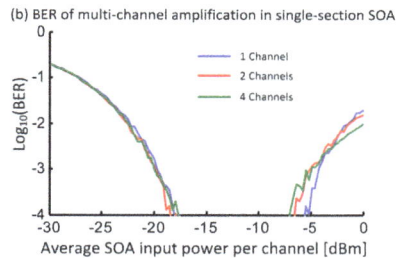

Figure 9. BER performance multi-channel PAM4 amplification performance in (**a**) multi-section SOA and (**b**) single-section SOA. Results are presented for one, two, and four co-propagating channels. At low input powers below −15 dBm, the ASE of the SOAs dominates and the performance is independent of the number of channels. The multi-section SOA performs better due to the lower effective NF. At high input powers above −10 dBm, there is a reduction in the useful dynamic range due to the increased crosstalk from the co-propagating channels.

5. Discussion and Conclusions

We showed that the BER performance of pre-amplified PAM4 signals can be modeled using simple, fundamental concepts to produce results that are in agreement with experimental results. Our results show that a multi-section SOA could improve the amplification dynamic range by over 3 dB through improved noise performance and through less gain saturation than an equivalent single-section SOA. Even though we have used one specific set of SOA parameters, the results presented would be applicable to SOAs with different parameters for the following reasons. Firstly, the NFs are typical of SOAs and the SOA noise depends mainly on the gain-inversion parameter [22], therefore knowledge of the NF will allow the signal to noise ratio (or/hence BER) at low input signal powers to be calculated. Secondly, for amplification purposes, SOAs with a moderate gain of about 13 dB gain to avoid excessive gain saturation that is limited by the SOA saturation power. Finally, the IPDR is maximized by employing SOAs with moderate gain of 13 dB to 15 dB. In this work, we solely focused on simulating SOAs with a moderate gain of 13 dB. SOAs with high unsaturated gain will have smaller IPDRs because of the smaller upper power limit at which the BER performance will be limited by the gain saturation. For multi-channel amplification, the BER performances for a single wavelength channel and for low-power multi-channels are in agreement with experiments. The true nature of the gain saturation, especially the wavelength dependence of the gain saturation, needs to be investigated further because there may be an optimal spectral placement of the signals with channel spacings of the order of 100 GHz that would allow for improved multi-channel performance when amplifying using SOAs and thus may enable greater routing flexibility in future datacom networks.

Acknowledgments: We acknowledge Science Foundation Ireland for funding this work (12/RC/2276, 14/IFB/2702). Seán P. Ó Dúill acknowledges fruitful discussions with René Bonk, Swen Koenig and Juerg Leuthold on quantifying SOA performance.

Author Contributions: This work was conceived by S.P.Ó.D. with suggestions from P.L. and L.P.B.; S.P.Ó.D. wrote the computer code to perform all of the simulations from which the results are presented here. All authors contributed to writing the paper.

Conflicts of Interest: The authors declare no conflict of interest. The founding sponsors had no role in the design of the study; in the collection, analyses, or interpretation of data; in the writing of the manuscript, and in the decision to publish the results.

References

1. Agrell, E.; Karlsson, M.; Chraplyvy, A.R.; Richardson, D.J.; Krummrich, P.M.; Winzer, P.; Roberts, K.; Fischer, J.K.; Savory, S.J.; Eggleton, B.J.; et al. Roadmap of optical communications. *IOP J. Opt.* **2016**, *18*. [CrossRef]
2. IEEE P802.3bs 200GbE & 400GbE Task Force. Available online: http://www.ieee802.org/3/bs/public/index.html (accessed on 2 September 2017).
3. NeoPhotonics Samples PAM4-Based 400G Pluggable CFP8 Transceiver, in Optical Connections News. 2017. Available online: http://opticalconnectionsnews.com/2017/02/neophotonics-samples-pam4-based-400g-pluggable-cfp8-transceiver/ (accessed on 2 September 2017).
4. Cheng, Q.; Ding, M.; Wonfor, A.; Wei, J.; Penty, R.V.; White, I.H. The feasibility of building a 64×64 port count SOA-based optical switch. Presented at the Photonics in Switching Conference, Florence, Italy, 22–25 September 2015.
5. Stabile, R.; Roht, A.; Williams, K.A. Monolithically integrated 8×8 space and wavelength selective cross-connect. *J. Lightwave Technol.* **2014**, *32*, 201–207. [CrossRef]
6. Anagnosti, M.; Caillaud, C.; Paret, J.-F.; Pommereau, F.; Glastre, G.; Blache, F.; Achouche, M. Record gain $\times$ bandwidth (6.1 THz) monolithically integrated SOA-UTC photoreceiver for 100-Gbit/s applications. In Proceedings of the 40th European Conference on Optical Communication (ECOC), Cannes, France, 21–25 September 2014.
7. Feris, B.D.; Gravey, P.; Morel, P.; Moulinard, M.-L.; Morvan, M.; Sharaiha, A. Dimensioning of 112G optical-packetswitching-based interconnects for energy-efficient data centers. *IEEE/OSA J. Opt. Commun. Netw.* **2017**, *9*, B124–B136. [CrossRef]
8. Bonk, R.; Huber, G.; Vallaitis, T.; Koenig, S.; Schmogrow, R.; Hillerkuss, D.; Brenot, R.; Lelarge, F.; Duan, G.-H.; Sygletos, S.; et al. Linear semiconductor optical amplifiers for amplification of advanced modulation formats. *Opt. Express* **2012**, *20*, 9657–9672. [CrossRef] [PubMed]
9. Koenig, S.; Bonk, R.; Schmuck, H.; Poehlmann, W.; Pfeiffer, T.; Koos, C.; Freude, W.; Leuthold, J. Amplification of advanced modulation formats with a semiconductor optical amplifier cascade. *Opt. Express* **2014**, *22*, 17854–17871. [CrossRef] [PubMed]
10. Lennox, R.; Carney, K.; Maldonado-Basilio, R.; Philippe, S.; Bradley, A.L.; Landais, P. Impact of bias current distribution on the noise figure and power saturation of a multicontact semiconductor optical amplifier. *Opt. Lett.* **2011**, *36*, 2521–2523. [CrossRef] [PubMed]
11. Carney, K.; Lennox, R.; Maldonado-Basilio, R.; Philippe, S.; Surre, F.; Bradley, L.; Landais, P. Method to improve the noise figure and saturation power in multi-contact semiconductor optical amplifiers: Simulation and experiment. *Opt. Express* **2013**, *21*, 7180–7195. [CrossRef] [PubMed]
12. Motawey, T.; Morel, P.; Sharaiha, A.; Brenot, R.; Verdier, A.; Guégan, M. Wideband gain MQW-SOA modeling and saturation power improvement in a tri-electrode configuration. *IEEE J. Lightwave Technol.* **2017**, *35*, 2003–2009. [CrossRef]
13. Kim, H.-S.; Choi, B.-S.; Kim, K.-S.; Kim, D.C.; Kwon, O.-K.; Oh, D.-K. Improvement of modulation bandwidth in multisection RSOA for colorless WDM-PON. *Opt. Express* **2009**, *17*, 16372–16378. [CrossRef] [PubMed]
14. Dúill, S.P.Ó.; Barry, L.P. Improved reduced models for single-pass and reflective semiconductor optical amplifiers. *Opt. Commun.* **2015**, *334*, 170–173. [CrossRef]
15. Way, W.; Chan, T.; Lebedev, A. Technical feasibility study of 56 Gb/s and 112 Gb/s PAM-4 transmission. Presented at the IEEE Meeting on P802.3bs Task Force 200 GbE and 400 GbE, Norfolk, VA, USA, 12–14 May 2014.
16. Kazovsky, L.; Benedetto, S.; Wilner, A. *Optical Fiber Communication Systems*; Artec House Inc.: Norwood, MA, USA, 1997; ISBN 0-89006-756-2.
17. Agrawal, G.P.; Olsson, N.A. Self-phase modulation and spectral broadening of optical pulses in semiconductor laser amplifiers. *IEEE J. Quant. Electron.* **1989**, *25*, 2297. [CrossRef]
18. Cassioli, D.; Scotti, S.; Mecozzi, A. A time-domain computer simulator of the nonlinear response of semiconductor optical amplifiers. *J. Quant. Electron.* **2000**, *36*, 1072–1080. [CrossRef]
19. Naimi, S.T.; Duill, S.O.; Barry, L.P. All optical wavelength conversion of Nyquist-WDM superchannels using FWM in SOAs. *IEEE J. Lightwave Technol.* **2016**, *33*, 3959–3967. [CrossRef]

Appl. Sci. **2017**, *7*, 908

20. Carney, K.; Lennox, R.; Watts, R.; Philippe, S.; Bradley, L.; Landais, P. Short pulse transmission characteristics in multi-contact SOA. In Proceedings of the 2012 14th International Conference on Transparent Optical Networks (ICTON), Coventry, UK, 2–5 July 2012.
21. Doi, Y. Applications of arrayed waveguide gratings for 100GbE-and-beyond datacom networks. Presented at the Photonics Networks and Devices, Boston, MA, USA, 27 June–1 July 2015.
22. Shimada, S.; Ishio, H. *Semiconductor Optical Amplifiers and Their Applications*; Wiley: Chichester, UK, 1994; ISBN 0471940054.

applied sciences

MDPI

Article

Bit- and Power-Loading—A Comparative Study on Maximizing the Capacity of RSOA Based Colorless DMT Transmitters

Simon Arega Gebrewold [1,2,*], Romain Bonjour [1], Romain Brenot [3], David Hillerkuss [1,4] and Juerg Leuthold [1]

[1] Institute of Electromagnetic Fields (IEF), ETH Zurich, 8092 Zurich, Switzerland; rbonjour@ethz.ch (R.B.); david.hillerkuss@huawei.com (D.H.); Juergleuthold@ethz.ch (J.L.)

[2] Now with RANOVUS GmbH, Nordostpark 07, 90411 Nuremberg, Germany

[3] Thales Research and Technology and CEA Leti, Route de Nozay, a Joint Lab of Alcatel-Lucent Bell Labs France, III-V Lab, 91460 Marcoussis, France; romain.brenot@3-5lab.fr

[4] Now with Huawei Technologies Duesseldorf GmbH, Optical & Quantum Laboratory, Riesstrasse 25-C3, 80992 Munich, Germany

* Correspondence: simon@ranovus.com; Tel.: +49-176-9093-9230

Received: 6 August 2017; Accepted: 4 September 2017; Published: 27 September 2017

Abstract: We present a comparative study of the capacity increase brought by bit- and power-loading discrete multi-tone (DMT) modulation for low-cost colorless transmitters. Three interesting reflective semiconductor optical amplifier (RSOA) based colorless transmitter configurations are compared: First, an amplified spontaneous emission (ASE) spectrum-sliced source; second, a self-seeded RSOA fiber cavity laser (FCL) and third, an externally seeded RSOA. With bit- and power-loaded DMT, we report record high line rates of 6.25, 20.1 and 30.7 Gbit/s and line rates of 4.17, 10.1 and 24.5 Gbit/s in a back-to-back and in a 25 km nonzero dispersion shifted fiber (NZDSF) transmission experiments for the three transmitter configurations, respectively. In all the experiments, BER (bit error ratios) below an FEC (forward error correction) limit of 7.5×10^{-3} were achieved.

Keywords: bit- and power-loading; colorless transmitter; discrete multi-tone (DMT); reflective semiconductor optical amplifier (RSOA); WDM-PON

1. Introduction

Maximizing the bit-rate of colorless, inexpensive and "poorly" performing transmitters has become a research topic on its own. This is of particular importance in wavelength division multiplexed (WDM) passive optical networks (PONs) that are considered a possible solution to cope with the capacity surge in access networks [1,2].

WDM-PONs are being considered for applications requiring bidirectional high capacity and secure links. WDM-PONs are therefore ideal solutions to serve fiber to the home and building (FTTH/B) users with high capacity demands and also ideal to cover the needs of mobile fronthaul (fiber to the antenna, FTTA) [3,4]. The latter is characterized by small cell sizes with dense base stations. If WDM-PON are to be mass deployed, large quantity of wavelength specific optical network units (ONUs) would be required, which highly complicates the inventory. Thus, the challenge is to develop ONU transmitters that allow the installation of identical devices across all subscribers. The solution are colorless transmitters that operate across a broad wavelength range, while their specific operation wavelength is controlled by an external factor [5–8]. Installing identical colorless transmitters across all ONUs consequently, reduces the number of specialized components—thus simplifying the inventory. In addition, relatively low-quality mass producible colorless sources will make WDM-PON more economical.

For low-cost colorless transmitters, the specific operating wavelength can be controlled by two main methods. A filter at a remote location can automatically choose the transmission wavelength of an ONU [9,10]. Alternatively, colorless operation can also be implemented by using an external seed laser placed at the central office (CO) to determine the upstream wavelength [11–16].

One promising colorless source is based on reflective semiconductor optical amplifiers (RSOAs). RSOAs are ideal sources as they offer a broad optical bandwidth at a low cost. There are three main architectures to integrate RSOAs in a WDM-PON as upstream (US) transmitters at the ONU. A theoretical and experimental comparison of the three architectures has been presented in [17].

- The simplest and most economical scheme is using RSOAs as amplified spontaneous emitters (ASE) source in a spectrum-sliced scheme, as shown in Figure 1a. The US signal is directly modulated onto the ASE of the RSOA. At the remote node (RN), a WDM filter slices the modulated ASE, passing only the assigned sub bands from each ONU to the central office (CO) [18,19]. The scheme is low-cost and simple, but so far only bit rates below 1 Gbit/s have been achieved for each wavelength, with this scheme. The limitations of this scheme are high filtering losses and susceptibility to chromatic dispersion due to a broad optical spectrum.
- Another cost-efficient source—however with improved performance—is the self-seeded RSOA fiber cavity laser (RSOA-FCLs) as suggested in [20,21] and depicted in Figure 1b. In this case, a mirror is placed at the RN behind the WDM filter. This establishes a feedback mechanism to the RSOA to form a Fabry-Perot laser resonator. The operating wavelength of each ONU's RSOA-FCL is determined by the WDM filter port it is connected to. This passive assignment of the emission wavelength simplifies the wavelength control. Thus far, RSOA-FCL have shown to transmit up to 10 Gbit/s [22,23].
- The highest performance—however, a relatively expensive scheme—is the externally seeded RSOA, shown in Figure 1c. Here, a bank of lasers, one for each ONU, is installed at the CO. The WDM filter at the RN redistributes the individual cw (continuous wave) laser lines to each ONU. At the ONUs the RSOAs amplify the seeding light and encode the US signal. Externally seeded RSOAs have already shown transmission beyond 20 Gbit/s with pulse amplitude modulation (PAM) [24] and more than 25 Gbit/s using discrete multi-tone (DMT) [14,25]. However, in both cases, controlled offset filtering by a WDM filter or an additional delay line interferometer was required, which makes the system sensitive to any wavelength drift. In addition, the extra lasers at the CO increases cost and energy consumption.

In these three schemes, the modulation bandwidth limitation of RSOAs is a challenge that has to be addressed to fulfill the high data rate requirements of WDM-PONs. There have been different approaches to alleviate this issue. One solution is to improve the speed of the RSOAs by optimizing the physical design of the device [26]. Initial implementations of RSOA, mainly in the externally seeded scheme, employed non-return-to-zero (NRZ) amplitude-shift keying (ASK) modulation with either electronic [27,28] or optical signal processing [28,29] or a combination of electronic and photonic signal processing [30] to demonstrate 10 Gbit/s. Return-to-zero (RZ) and NRZ are compared in [31]. Advanced modulation formats have also been demonstrated for higher spectral efficiency with electronic and optical equalization in [12]. Polar RZ PAM4 with delay interferometer optical signal processing has achieved 20 Gbit/s [32]. Recently, multicarrier schemes such as adaptive orthogonal frequency division multiplexing (Adaptive-OFDM) [33] have been implemented for improved performance of the externally seeded scheme. Adaptive OFDM [25] and bit- and power-loaded DMT [14], both with optical equalization have already demonstrated beyond 25 Gbit/s transmission.

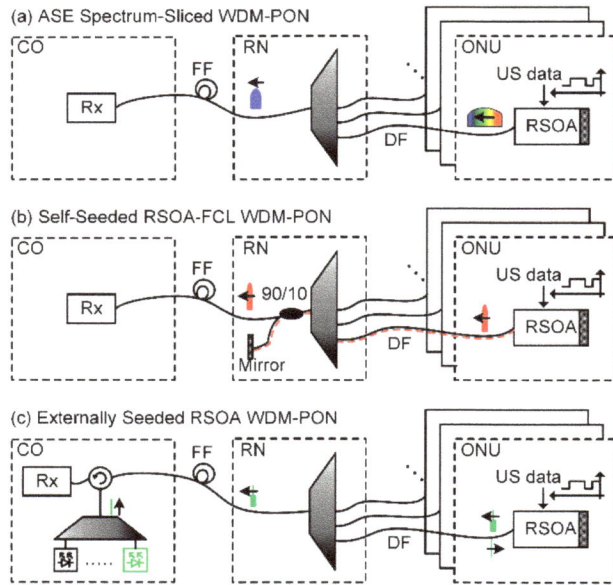

Figure 1. Wavelength division multiplexed passive optical network (WDM-PON) topologies with reflective semiconductor optical amplifier (RSOA) based colorless upstream (US) transmitters: (**a**) Amplified spontaneous emitter (ASE) spectrum-sliced transmitter: the US signal is directly modulated on to the ASE of the RSOA. The WDM filter at the remote node (RN) passes only the assigned sub band from each RSOA at the optical network units (ONUs); (**b**) Self-seeded RSOA-FCL (fiber cavity laser): a Fabry-Perot cavity indicated by the red dashed lines is formed between the RSOA and a mirror at the RN. The cavity is embedded in the distribution fiber (DF) of the PON; (**c**) Externally seeded RSOA: a laser at the central office (CO) seeds the RSOA at the ONU. The RSOA encodes the US signal onto the seed laser line. In all case, the ONUs are connected by the DF to the RN. The RN consists of the WDM filter and it is connected to the CO through the feeder fiber (FF).

In general, DMT has been used to maximize capacities of bandwidth limited intensity modulated direct detected (IM-DD) systems [25,34–36]. DMT is a type of OFDM with real-valued output signals that can be easily detected with inexpensive direct detection receivers. As a multicarrier format, DMT also brings the flexibility of adapting the modulation format and power of any individual subcarrier to cope with the frequency dependence of the channel signal-to-noise ratio (SNR).

In this paper, we employ bit- and power-loaded DMT to show that the ASE spectrum-sliced scheme, the self-seeded RSOA-FCL, and the externally seeded RSOA configurations are viable US transmitters. The results are an expansion of our recent work [37], where only the maximized bit rates are presented. In this work, the SNR offered by each sources has been measured and compared. It is found that the SNR improves by more than 5 dB when switching from an ASE spectrum-sliced scheme to a self-seeded RSOA-FCL, and it improves by another ~5 dB when switching from a self-seeded RSOA-FCL to an externally seeded RSOA. Finally, transmission experiments have been performed. We demonstrate 6.25, 20.1, and 30.7 Gbit/s in back-to-back (BtB) and 4.17, 10.1 and 24.5 Gbit/s over 25 km NZDSF (non-zero dispersion shifted fiber) for the ASE spectrum-sliced, RSOA-FCL and externally seeded RSOA sources, respectively. No offset filtering was needed in any of these experiments [14,25].

The organization of the paper is as follows. First, we discuss the steps for generating and analyzing a DMT signal. In Section 3, we present the experimental setups. In Section 4, we measure the SNR of the RSOA-FCL by varying WDM filter bandwidth, cavity length and bias current. We compare these results with the optimum SNR of the ASE spectrum-sliced and externally seeded RSOA schemes. Finally we present the results of bit- and power-loaded DMT transmission experiments.

2. DMT Signal Generation and Reception

DMT is a multi-carrier transmission technique with real-valued orthogonal subcarriers. The signal processing flow charts for a DMT transmitter and receiver are given in Figure 2. A pseudorandom bit sequence (PRBS) is generated to test the transmitter. The bits are then rearranged from serial into parallel according to the number of subcarriers and the information content in the respective subcarriers. Each parallel bit stream is then mapped onto its individual constellation (e.g., quadrature amplitude modulation—QAM). DMT gives us the flexibility to encode a different modulation format onto every subcarrier. Therefore, depending on the available SNR at each subcarrier the number of bits per symbol can be adapted to achieve the highest possible throughput, a feature called bit-loading. In our experiments, the modulation schemes ranged from binary phase shift keying (BPSK with 1 bit/symbol) to 64QAM with 6 bit/symbol. The constellation diagrams of the modulation formats are shown in Figure 3. After bit mapping, the power of each subcarrier is optimized in a power-loading stage depending on the SNR profile, see Figure 2. This is done such that all subcarriers will end up with a relatively similar bit error ratio (BER) performance even if they have different modulation formats. The adaptive bit-and power-loading algorithm is based on Chow's method discussed in [38]. In a next step, an inverse fast Fourier transform (IFFT) with Hermitian symmetry is performed to obtain the real valued DMT symbols. A cyclic prefix is added to each DMT symbol before being serialized and loaded into an arbitrary waveform generator (AWG). The AWG converts the DMT signal to the analog domain before it is fed to an RF amplifier and encoded to the optical domain by means of an RSOA.

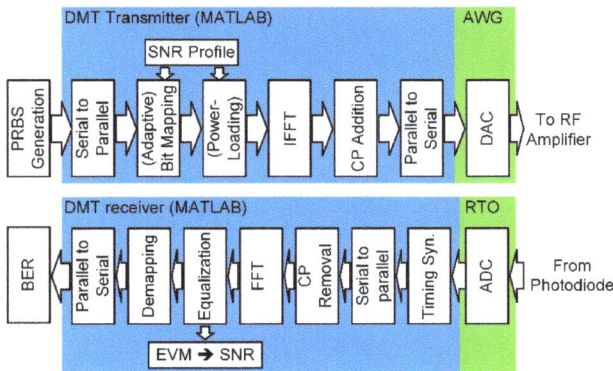

Figure 2. Discrete multi-tone (DMT) signal generation and reception flowchart. A pseudorandom bit sequence (PRBS) is generated, which is then rearranged into a parallel bit-stream. Subsequently, adaptive bit mapping and power-loading are implemented. Inverse fast Fourier transform (IFFT) is then used to encode the quadrature amplitude modulation (QAM) symbols on the DMT subcarriers. Cyclic prefix (CP) is added on the DMT symbols before the signal is converted back to serial. The serial data is loaded onto an arbitrary waveform generator (AWG) which converts the digital signal to analog using an inbuilt digital to analog converter (DAC). On the receiver side the signal is detected by a photodiode and sampled by a real-time oscilloscope (RTO).Next, timing synchronization is performed to recover the DMT clock. Then the serial signal is converted into a parallel data stream. After CP removal, FFT is used to decode the DMT subcarriers. Frequency and phase equalization of the received QAM symbols is performed. Here, we measure the error vector magnitude (EVM) and signal-to-noise ratio (SNR). Then, the symbols on each subcarrier are demapped to recover the transmitted PRBS for bit error ratio (BER) measurements.

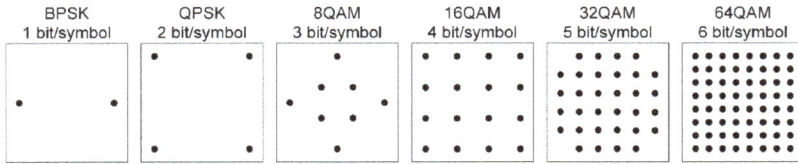

Figure 3. Constellation diagrams of modulation formats used for the bit- and power-loaded DMT signal generation. Binary phase shift keying (BPSK) encodes 1 bit/symbol. Quadrature phase shift keying (QPSK) encodes 2 bit/symbol. The quadrature amplitude modulation (QAM) formats: 8QAM, 16QAM, 32QAM, and 64QAM encode 3, 4, 5 and 6 bit/symbol, respectively.

In the receiver, a direct detection scheme is used to detect the optical signals. The electrical signals are then fed into a real-time oscilloscope (RTO). The RTO effectively digitizes the signal with a resolution of ~5.8 bits. In a first step, timing synchronization is performed using the non-data aided DMT timing estimation technique discussed in [39]. Subsequently, the signal is parallelized and the cyclic prefix is removed.The subcarriers within the electrical signal are then derived by the fast Fourier transform (FFT). To properly decode the bits within a subcarrier, blind phase and frequency offset estimation methods are applied on every subcarrier [40]. The error vector magnitude (EVM) of the QAM symbols on each subcarrier is calculated and converted into SNR [41]. Then, the QAM symbols are de-mapped to extract the received bit pattern. The BER is measured by comparing the transmitted and received bit patterns on each subcarrier. The overall BER is also measured by comparing the total transmitted and received bits.

3. Experimental Setups

In our experiments, we investigated the three colorless US transmitters in WDM-PON scenarios, see Figure 4. For all experiments, the transmitted DMT signal is generated offline with MATLAB and converted to an analog waveform by an AWG (Keysight M8195A—60 GSa/s, Santa Rosa, CA, USA). The analog signal is amplified and a DC bias current is added before being fed to an RSOA. In all cases, a 1 mm long InGaAsP quantum well C-band RSOA (provided by III-V labs, 10G-RSOA-11-S, Palaiseau, France) is placed at the ONU. The distribution fiber (DF) links the ONU to the RN. At the RN, a flat top optical band-pass filter (OBPF) emulates the WDM filter. RN and CO are connected by the feeder fiber (FF).

The three implementations of the RSOA based transmission links are shown in Figure 4 and described in more detailed now.

Figure 4a shows the setup for the ASE spectrum-sliced scheme. In this scheme, the current of an ROSA in the ONU is modulated with the data. This encodes the information onto the ASE output power of the RSOA. At the remote note, an OBPF (tuned to the emission center of the RSOA, which is at 1537.92 nm) band limits the modulated ASE-signal to the assigned wavelength channel. The filtered signal is then sent to the CO. The filter bandwidth has been chosen to be identical to the filter bandwidth of the other schemes.

Figure 4b shows the setup for the self-seeded RSOA-FCL. Here, a Faraday rotator (FR) is placed at ONU after the RSOA. In addition, a Faraday rotator mirror (FRM) is placed at the RN after a coupler behind the WDM filter. The FRM and the RSOA reflective facet form the Fabry-Perot resonator cavity embedded in the DF. In case of RSOA with high polarization dependent gain (HPDG), the FR and FRM cancel out the birefringence effect in the cavity and maintain a constant polarization at the RSOA input [42]. The coupler at the RN has a coupling ratio of 90/10, where 90% of the power is kept in the cavity and 10% is transmitted to the CO. The large amount of the power kept in the cavity is required to overcome cavity losses and thereby to achieve a net gain and with this lasing. In addition, the high power in the cavity maintains the RSOA in the nonlinear operation region for better noise and residual modulation suppression [17,43].

Figure 4c shows the experimental setup for the externally seeded RSOA scheme. Here, a laser source (emitting at 1537.92 nm, similar to the other setups) at the CO seeds the RSOA at the ONU through the fiber link. The seeding power entering the RSOA has been optimized for the highest SNR. The RSOA now only serves as a modulator and amplifier for the respective laser wavelength.

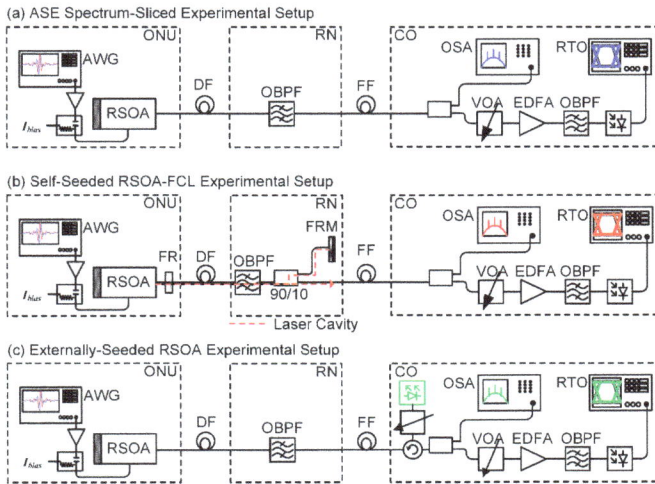

Figure 4. Experimental setups of three RSOA based colorless transmitter schemes in a WDM-PON scenario: (**a**) ASE spectrum-sliced, (**b**) self-seeded RSOA-FCL, and (**c**) externally seeded RSOA. At the optical network units (ONUs), the upstream signal from the arbitrary waveform generator (AWG) is amplified and added to a bias current to drive the RSOA. The ONU and the remote node (RN) are connected by the distribution fiber (DF). At the RN, an optical bandpass filter (OBPF) mimics the WDM filter. In the case of the RSOA-FCL in (**b**), the Faraday rotator mirror (FRM) is placed behind the filter and a coupler to form the Fabrey-Perot cavity, shown by the red dashed line. For the externally-seeded RSOA in (**c**), the seeding cw laser is place at the central office (CO). On the receiver side at the CO, pre-amplification and direct detection are performed. A real-time oscilloscope (RTO) records the electrical signal after the photodiode. An optical spectrum analyzer (OSA) is used to monitor the optical signal at the CO. EDFA: Erbium doped fiber amplifier, VOA: variable optical attenuator.

In the three schemes, we implement only a single optical channel measurement due to lack of equipment, however more than 16 channels have been demonstrated [2,21]. In all cases, the link budget between the ONU and the CO is ~12 dB (i.e., combination of all insertion losses of components in the link between ONU and CO). At the CO, a direct detection receiver detects the signals from the three transmitters. Here, a 90/10 coupler sends 10% of the power to an optical spectrum analyzer (OSA) to monitor the signal. A variable optical attenuator (VOA) is used to control the power entering the preamplifier Erbium doped fiber amplifier (EDFA). A pre-amplified direct detection receiver is used here to guarantee that the signal quality measured here is as much as possible independent of the receiver scheme. A 5 nm flat top OBPF suppresses the out of band ASE noise from the EDFA. The optical signal is finally detected by a photodiode (PIN, responsivity 0.6 A/W), which is directly connected to a RTO (DSOX96204Q—20 GHz bandwidth and 40 GSa/s). The oscilloscope records the signal for offline post-processing with MATLAB as outlined in Figure 2.

4. SNR Characterization

In this section, we present the measured SNRs versus subcarrier frequency offered by the three transmitter schemes. Before presenting the results, we study the performance of the RSOA itself by measuring the electrical modulation bandwidth. In a second step, we then focus on the self-seeded RSOA-FCL. We study the impact of the WDM filter bandwidth in the RN, the fiber cavity length and the bias current on the SNR offered by the RSOA-FCL. Once the optimum operation parameters and limitations of the RSOA-FCL have been understood, we adjust the operation points of the ASE spectrum-sliced and the externally seeded RSOA. So for instance we set the WDM filter bandwidth of the ASE spectrum-sliced source to be identical with the RSOA-FCL. The input power for the externally seeded transmitters is also set for maximum performance. Here we only give the operation points that have been selected while one can find more details on finding the ideal operation points of an ASE spectrum-sliced transmitter in [44,45] and the externally seeded transmitter [25,46].

4.1. Electrical Bandwidth of the RSOA

We start our investigation by measuring the small signal modulation frequency response of the RSOA itself. We chose the externally seeded configuration to exclude the impact of dispersion and the external cavity. The frequency response for 50–120 mA bias current and at an input power of −3 dBm are plotted in Figure 5, which shows that the modulation response improves with higher bias current. At 100 mA, the RSOA exhibits −3 and −6 dB bandwidths of 2.4 and 4.55 GHz, respectively.

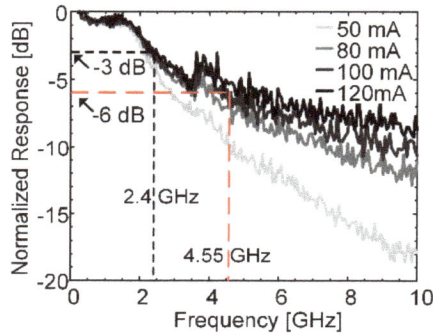

Figure 5. Measured small signal modulation frequency response of an RSOA at 50–120 mA bias currents. The modulation bandwidth increases with bias current. At 100 mA, the RSOA has −3 and −6 dB bandwidths of 2.4 and 4.55 GHz.

4.2. RSOA-FCL Performance

Here, we perform three studies to analyze the performance of the RSOA-FCL: First, we investigate the impact of the filter bandwidth in the cavity. Second, we study the SNR dependence on cavity length and bias current. Last, we explore the transmission performance for different cavity lengths at optimized operation conditions. In all cases, we use the SNR over the modulation frequency as a measure for the link performance and for comparing parameters.

The SNR over the modulation frequency is measured according to the following procedure. We transmit DMT signals and evaluate the SNR on each subcarrier at the receiver. Here, the modulation format and power of each subcarrier is identical. We generate a DMT signal having 256 subcarriers with QPSK as modulation format and equal electrical power for all subcarriers. A PRBS 11 is used as the transmitted bit sequence. The total symbol rate is 10 GBd. The IFFT size is 512 and a 2.5% cyclic prefix is added to each DMT symbol.

The signal is generated offline with MATLAB. An AWG generates the analog signal with 0.5 Vpp (peak-to-peak voltage). The output of the RF amplifier was optimized to ~3.5 Vpp. A bias tee combines the amplified signal with the specific bias current to drive the RSOA. On the receiver side, an RTO records the signal detected by a photodetector. Subsequently, the signal is demodulated in an offline digital signal processing step. The measured EVM of the subcarriers is used to calculate the SNR of each subcarrier. This way, we can determine the frequency resolved SNR of the transmission link across the whole modulation bandwidth. This allows us to determine the usable bandwidth of the transmission link. The useable bandwidth is the frequency band, within which the SNR is high enough to still allow for a BER below 10^{-3} with BPSK modulation. Theoretically, a minimum SNR of 6.7 dB is required.

First, we investigated the impact of the WDM filter bandwidth in the RN. We compared OBPF bandwidths of 0.6 nm and 2 nm for three cavity lengths of 15 m, 115 m and 1 km standard single mode fiber (SSMF). The bias current was maintained at 100 mA. Figure 6 shows the SNR over the modulation frequency for both cases. The SNR is higher for the filter with wider bandwidth for all cavity lengths. For a 0.6 nm filter, the 15 and 115 m cavities have a peak SNRs of ~14 dB. The maximum SNR obtained for the 1 km cavity with 0.6 nm is only ~12 dB. The usable bandwidth of the 15, 115 m and 1 km cavities with 0.6 nm filter was ~4.6, 4.5 and 2.7 GHz. When a 2 nm filter has been used in the receiver one finds SNRs that are 5 dB higher. The usable bandwidth then increases by ~40% to ~6.7, ~6.4 and ~4.5 GHz. The results are summarized in Table 1.

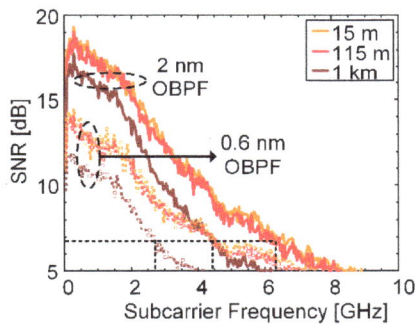

Figure 6. Measured SNR comparison of self-seeded RSOA-FCL transmitter with WDM filter bandwidths of 0.6 and 2 nm in RN. The dotted and solid lines show the measured SNR for the 0.6 and 2 nm filters, respectively. For both cases, three different cavity lengths, 15 m, 115 m and 1 km were used. 0.6 nm filter cavity has peak SNR which is ~5 dB lower than that of the 2 nm one.

Table 1. Summary of results to compare the peak signal-to-noise ratio (SNR) and usable SNR bandwidth for the RSOA-FCL (reflective semiconductor optical amplifier fiber cavity laser) with 0.6 nm or 2 nm filter bandwidths. For both filters, the fiber cavity length varied between 15 m, 115 m and 1 km. OBPF: optical band-pass filter.

RSOA-FCL Cavity Length	OBPF −3 dB Bandwidth	Peak SNR (dB)	Usable SNR Bandwidth (GHz)
15 m	0.6 nm	14.3	4.6
	2 nm	19.45	6.7
115 m	0.6 nm	14.1	4.5
	2 nm	19.5	6.4
1 km	0.6 nm	11.9	2.7
	2 nm	17.7	4.5

One explanation for the SNR degradation with narrower filters would be that a narrower filter leads to higher losses [47]. For example, at 100 mA, the reflected power into the RSOA was ~−6 dBm and ~−2 dBm for the 0.6 and 2 nm filters, respectively. Thus, higher losses means that the RSOA is less saturated. Therefore, the RSOA will be unable to efficiently suppress the residual modulation and noise. As a consequence the RIN will be higher and the SNR is degraded [47]. The second explanation based on [48] is that in case of highly multimode sources, increasing the number of modes (in this case wider filter) will distribute the noise over the multiple modes thus lowering the noise spectral density and RIN. Subsequently, we continue the investigation with the 2 nm filter only.

Next, we further investigated the impact of the distribution fiber cavity length and the bias current on the SNR of the self-seeded RSOA-FCL. We considered cavity lengths with 15 m, 115 m and 1 km SSMF to mimic WDM-PONs with distribution fibers in the range of tens of meters up to a few kilometers, for example in FTTA [49]. The RSOA bias current was set at 80, 100, 110 and 120 mA. Figure 7a,b show similar performance for cavities with 15 m and 115 m length. The insets in Figure 7a,b which are the transmitted (black curve) and the received (red curve) DMT electrical spectrums also show similar characteristics. The similar performance originates from negligible chromatic dispersion for the two lengths. In both cases, the frequency dependent SNR increases with the bias current, which is related to the improved RSOA frequency response for higher bias currents as shown in Figure 5. The SNR degradation for higher modulation frequencies is due to the bandwidth limitations of the RSOA as discussed in Figure 5. At 110 mA, both achieve maximum SNR of ~19 dB. The usable bandwidth where the SNR is still sufficiently high to receive a BPSK signal is ~6.7 and ~6.4 GHz for cavity lengths of 15 m and 115 m, respectively.

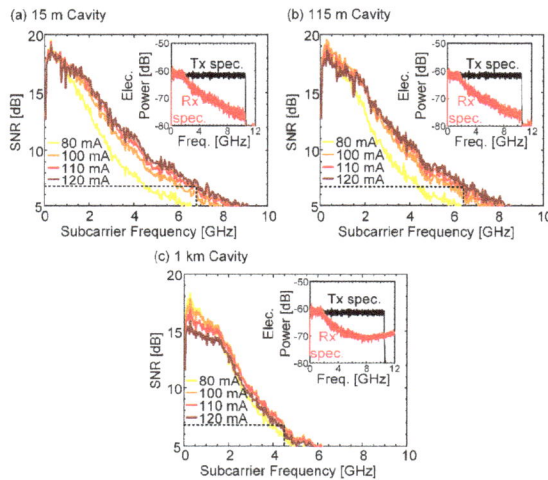

Figure 7. Measured SNR versus subcarrier frequency of self-seeded RSOA-FCLs operated at 80, 100, 110 and 120 mA bias currents for cavity lengths of (**a**) 15 m, (**b**) 115 m and (**c**) 1 km. The plots indicate that the SNR bandwidth generally increases with the bias current. The 1 km cavity shown in (**c**) has a reduced SNR peak and bandwidth when compared to the shorter cavities. The dashed line indicates the minimum SNR of 6.7 dB required to transmit BPSK, thus, it determines the usable bandwidth. In each plot, the insets show a sample electrical domain DMT spectra. The black and the red curves show the transmitted and received DMT electrical spectrum, respectively, at 110 mA. The received DMT spectrum for the 1 km cavity RSOA-FCL shown in plot (**c**) inset has a higher noise level compared to that of the 15 and 115 m cavities spectra shown in insets of (**a**,**b**).

In Figure 7c we have plotted the SNR for the longer 1 km SSMF cavity. It is way lower than those in Figure 7a,b. This can be understood by the fact that total accumulated chromatic dispersion is larger. Since the RSOA-FCL is highly multi-mode, chromatic dispersion leads to considerable mode partition noise. This noise increases the RIN [50,51], causing a SNR degradation. A bit higher loss of the 1 km fiber cavity makes only a minor contribution to the SNR increase. At 110 mA, the maximum SNR is ~17 dB only and the usable bandwidth is now ~4.5 GHz. In contrast to the RSOA-FCL with shorter cavity lengths, the SNR in the 1 km cavity is not increasing with bias current. We expect that this originates from the strong chirp during direct modulation at higher bias currents. The chirp causes spectral broadening which increases chromatic dispersion penalty. This induces an additional SNR degradation for higher bias currents [52]. This is supported by the large noise level in the received electrical DMT spectrum of the 1 km RSOA-FCL that can be seen in the inset of Figure 7c (red curve). However, the negative impact of the dispersive cavity on the SNR can be avoided by using O-band RSOAs. In the O-band where chromatic dispersion is minimum, RSOA-FCL cavity could be extended beyond 1 km [22,53].

Finally, we investigated the SNR after transmission over 25 km NZDSF (chromatic dispersion: ~5 ps/nm·km). This fiber is used to emulate a low dispersion PON fiber—similar to what one might expect if the system were operated in the O-Band. The probed SNR for the three cavity lengths for both a BtB and a 25 km transmission are plotted in Figure 8. When comparing the BtB and 25 km SNRs, the maximum SNR values for the respective cavity lengths did not show major change. Rather, the usable SNR bandwidth was reduced to ~3.3, 3.2 and 2.8 GHz for the 15 m, 115 m and 1 km cavities, respectively. The results are summarized in Table 2. This shows a 40–50% bandwidth reduction compared to the BtB case. The finite signal bandwidth of the directly modulated RSOA results in chirp after propagation that induces power fading and bandwidth reduction. In addition, such highly multi-mode source suffers from mode partition noise induced by the NZDSF dispersion which increases the RIN.

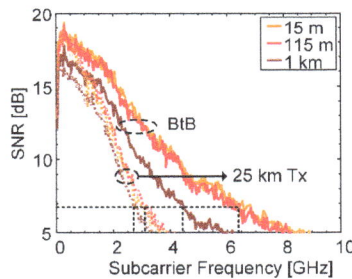

Figure 8. Measured SNR to compare the impact of transmission fiber on the performance of an RSOA-FCL: The solid and dotted lines show the measured SNR for BtB (back-to-back) and after 25 km non-zero dispersion shifted fiber (NZDSF) transmission, respectively. In both cases, the results are for 15 m, 115 m and 1 km fiber cavity lengths. The figure shows that the SNR bandwidth narrows by up to 50% after transmission over 25 km fiber.

Table 2. Summary of results for the SNR performance of the RSOA-FCL with 15 m, 115 m and 1 km cavities for back-to-back (BtB) and 25 km non-zero dispersion shifted fiber (NZDSF) transmission.

RSOA-FCL Cavity Length	Tx	Peak SNR (dB)	Usable SNR Bandwidth (GHz)
15 m	BtB	19.45	6.7
	25 km NZDSF	19.01	3.2
115 m	BtB	19.5	6.4
	25 km NZDSF	19.2	3.2
1 km	BtB	17.7	4.5
	25 km NZDSF	17.1	2.8

4.3. Performance Comparison of the Three Transmitter Schemes

Here, we compared the SNR performance of the ASE spectrum-sliced with the self-seeded RSOA-FCL and the externally seeded RSOA sources. For this experiment, the ASE spectrum-sliced source was equipped with the 2 nm filter, similar to that of the RSOA-FCL. For the RSOA-FCL, we took the 15 m and 1 km cavities, which offered the best and the worst performance, respectively. Both

RSOA-FCLs were equipped with a 2 nm filter. The RSOA input and output power for the externally seeded scheme was −3 dBm, and ~6 dBm, respectively. The bias current for all experiments was set to 100 mA. Figure 9a,b show the measured SNR values for BtB and transmission over 25 km NZDSF, respectively. It can be seen that the externally seeded RSOA and the self-seeded RSOA roughly offer a 10 dB and a 5 dB higher SNR over the ASE spectrum-sliced source across the whole spectral band. The self-seeded RSOA-FCL with the shorter cavity outperforms the longer cavity FCL by about 1–2 dB across the spectral band. And while the externally seeded RSOA exhibited the highest SNR bandwidth of ~9.6 GHz for BtB, the 15 m and 1 km long RSOA-FCLs offer useable bandwidths of 6.7 and 4.5 GHz and the ASE spectrum-sliced scheme features a usable bandwidth of only 2.5 GHz.

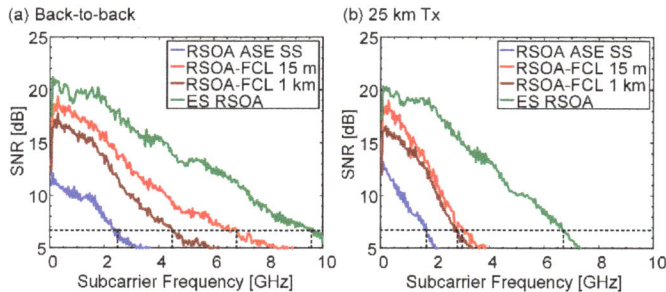

Figure 9. Comparison of the measured SNR for the three colorless transmitter schemes: RSOA ASE spectrum-sliced (SS), self-seeded RSOA-FCL and externally seeded (ES) RSOA. (**a**) Shows the BtB (back-to-back) SNR measurement. The externally seeded RSOA (green line) outperforms the RSOA-FCL (red and brown) and the ASE spectrum-sliced scheme (blue) by ~10 and ~5 dB SNR across the whole frequency range, respectively; (**b**) Shows the SNR after 25 km NZDSF transmission. The SNR bandwidth is reduced due to dispersion of the transmission fiber. In general, the RSOA-FCL offers performance midway between an ASE spectrum sliced and an externally seeded RSOA.

When transmitting the signals over 25 km NZDSF fiber, the usable SNR bandwidth reduced to ~1.6 GHz for the ASE spectrum-sliced source. The RSOA-FCL declines to a usable bandwidths of 3 and 2.8 GHz for the 15 m and 1 km long cavities. And the externally seeded RSOA still offers a 6.7 GHz bandwidth with sufficient SNR for transmission. The source of the SNR degradation with transmission is the interplay of the chirp (induced by the direct modulation of the RSOA) and the dispersion of the transmission fiber. The narrower usable bandwidth recorded after transmission will influence the maximum achievable bit rate.

We attribute the higher performance observed for the RSOA-FCL when compared to the ASE spectrum-sliced source to three main factors. First, the RSOA-FCL has a significantly narrower spectrum, which significantly reduces the impact of chromatic dispersion [17,50]. Second, the RSOA-FCL has a larger output power. For example, at 100 mA, the RSOA-FCL has an output power of −3.4 dBm while the ASE spectrum sliced scheme emits −11.8 dBm. Third, the saturated RSOA in the cavity nonlinearly suppress the amplitude noise in the lasing cavity, which lowers the RIN. These three contribute to the improve SNR of the RSOA-FCL compared to the ASE spectrum sliced source [17].

5. Bit- and Power-Loaded DMT for Highest Bit Rates

In this section, bit- and power-loaded DMT transmission is used to maximize the data throughput for any of the three bandwidth limited sources. We proceed in two steps. We again first study the impact of the cavity length on the RSOA-FCL and the achievable data throughput in more detail. Only when we know the optimum operation conditions of the RSOA-FCL, we investigate and compare the maximum capacity of all three schemes.

5.1. Maximizing the Capacity of the RSOA-FCL

In our experiments, we performed bit- and power-loading based on Chows algorithm [38]. Chow's algorithm can briefly be described as follows. It takes in the SNR characterization of a transmitter at the subcarrier frequencies of the DMT signal. It also uses the SNR requirement of all advanced modulation formats (see Figure 3) to achieve a predetermined BER. Then, it disregards subcarriers with the lowest SNR that cannot support transmission at the predetermined BER for even the simplest format. The power of such unusable subcarriers is then transferred to usable subcarriers. For subcarrier with non-integer bit per symbol, the bit number is rounded off to the nearest integer, which requires the power to be re-adapted accordingly. This power adaptation step (power-loading) helps to achieve similar BER for all subcarriers.

First, we performed bit- and power-loaded DMT transmission for the RSOA-FCL with 15 m, 115 m and 1 km cavities. The signal was optimized for a BtB experiment and for transmission over 25 km NZDSF. Figure 10a–c show the SNR, and the bit- as well as the power-loading patterns of the DMT and the measured per-subcarrier BER for 15 m, 115 m and 1 km long RSOA-FCLs, respectively. The 15 and 115 m cavities can support up to 5 bit/symbol (32QAM) for low frequency subcarriers. In case of the 1 km RSOA-FCL, the highest order modulation format is 16QAM, with 4 bits/symbol. The power-loading pattern is arranged to guarantee a minimum SNR difference for all subcarriers with the same modulation format, so that a similar BER performance can be obtained. The solid lines in the BER plots indicate the overall BER, i.e., the total BER across all subcarriers. It can be seen that the overall BER is below 9.29×10^{-3} in all instances, which is the limit for an FEC (forward error correction) with 12.5% overhead [54].

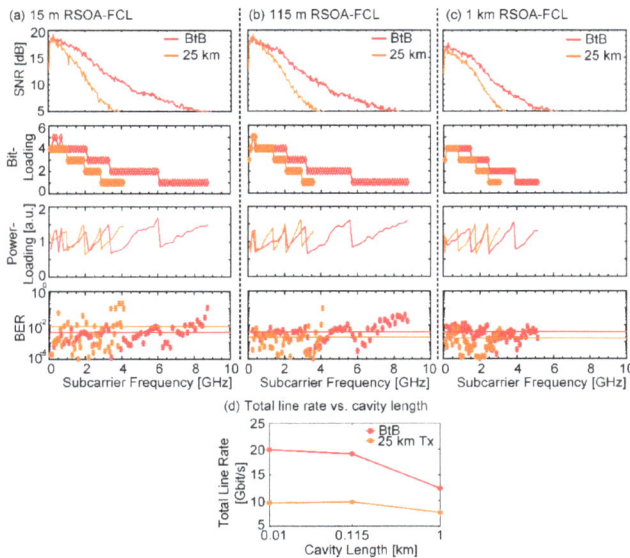

Figure 10. Measured SNR, bit- and power-loading pattern, and bit error ratio (BER) versus sub-carrier frequencies and the achieved total line rate for different RSOA-FCL cavity lengths: Plots (**a**–**c**) show the results for 15, 115 m and 1 km long cavities, respectively. The red and orange colors indicate BtB (back-to-back) and 25 km NZDSF transmission measurements. In all cases, the overall BER (solid lines in the BER plots) are kept below the FEC limit of 9.29×10^{-3} [53]; (**d**) Plots the measured total line rate for 15, 115 m and 1 km long cavities, both for BtB and after 25 km transmission. In a BtB measurement, the RSOA-FCL offers up to 20 Gbit/s line rates. The line-rate is still 10 Gbit/s when transmitting signals over 25 km.

Figure 10d shows the maximum line rates obtained with RSOA-FCL cavity lengths of 15 m, 115 m and 1 km length for a BtB and a transmission over 25 km NZDSF. Maximum capacities are obtained for short FCL cavities of 15 and 115 m length. The capacities are between 20 and 10 Gbit/s depending on the transmission distance. For longer cavity lengths the capacity dropped to 12.4 Gbit/s due to a reduced SNR and a limited bandwidth for the BtB experiment and 7 Gbit/s could at best be transmitted over 25 km.

5.2. Comparison of Maximizing Capacity

Second, we compared the maximum capacity achievable with bit- and power-loaded ASE spectrum-sliced, two types of self-seeded FCL RSOAs and externally seeded RSOAs.

Figure 11a,b show BtB and 25 km DMT transmission results, respectively. For low subcarrier frequencies, the ASE spectrum-sliced source supported up to 3 bit/symbol (8QAM), the RSOA-FCL up to 5 bit/symbol (32QAM), and the externally seeded RSOA up to 6 bit/symbol (64QAM). The power-loading pattern shown in Figure 11a,b led to relatively similar SNRs for subcarriers with the same modulation format. As a result, the BER for different subcarriers was relatively constant over the modulation frequency. The solid lines show the overall BER. For all schemes, the overall BERs were kept below 7.5×10^{-3} (10% overhead FEC) [54].

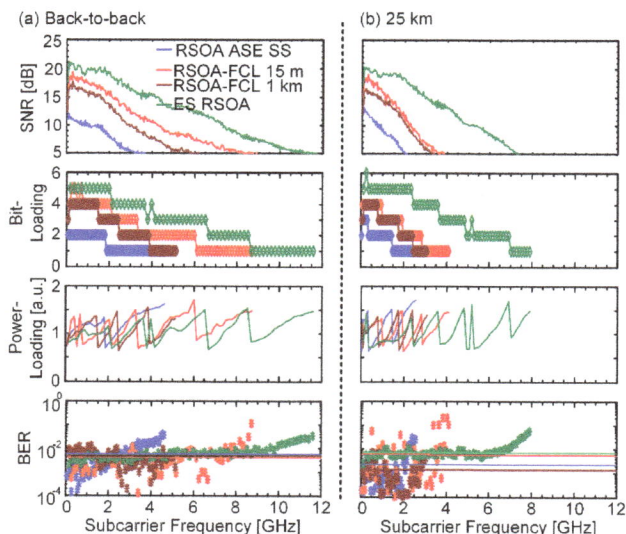

Figure 11. Measured SNR, bit- and power-loading pattern and per-subcarrier BER for the three transmitter schemes: RSOA ASE spectrum-sliced (SS) (blue), self-seeded RSOA-FCL (red and brown) and externally seeded (ES) RSOA (green). (**a**) and (**b**) show results for back to back case and 25 km transmission. The ASE spectrum sliced scheme can support modulation formats with up to 3 bits/symbol while RSOA-FCL and externally seeded RSOA can encode up to 6 bits/symbol for low subcarrier frequencies. The solid line in the bottom plot indicates the overall BER that can be achieved with either of the three transmission schemes. All bit-error ratios are within the FEC (forward error correctio) limit of 7.5×10^{-3}.

Table 3 summarizes the achieved maximum line rates for the three transmitter schemes and the corresponding overall BER. In a BtB experiment the ASE spectrum-sliced, the RSOA-FCL and the externally seeded sources were delivering capacities up to 6.25, 20.1 and 30.7 Gbit/s, respectively. For a link with 25 km the sources could still deliver maximum capacities of 4.2, 10.1 and 24.5 Gbit/s.

It is interesting to note that the chirp-dispersion induced usable bandwidth reduction results in considerable capacity reduction, especially in the case of the ASE spectrum-sliced and RSOA-FCL transmitters. This can be explained by the relatively wider linewidth of the two sources, thus incurring higher dispersion penalty. However, O-band RSOAs [53], can be used to lower dispersion penalty and avoid the need for NZDSF. O-band has already been used for upstream transmission in legacy PONs to relax the requirement on inexpensive poor sources [1,55].

Table 3. Summary of achieved line rate and the respective bit error ratio (BER) for the three transmitter schemes. The results are for a BtB (back-to-back) and after 25 km NZDSF transmission. ASE: amplified spontaneous emitters.

	ASE Spectrum-Sliced		RSOA-FCL (15 m)		RSOA-FCL (1 km)		Externally Seeded RSOA	
	Line Rate (Gbit/s)	BER	Line Rate (Gbit/s)	BER	Line Rate (Gbit/s)	BER	Line Rate (Gbit/s)	BER
BtB	6.25	5.8×10^{-3}	20.1	5.8×10^{-3}	12.4	5.8×10^{-3}	30.7	5.7×10^{-3}
25 km NZDSF	4.2	2.3×10^{-3}	10.1	4×10^{-3}	7	5.8×10^{-3}	24.5	7.1×10^{-3}

6. Conclusions

We show how transmission capacities of low-cost low-quality sources can be optimized for maximum capacities by means of bit- and power-loaded DMT transmission. Record high transmission data rates for an ASE spectrum-sliced, self-seeded RSOA-FCL and externally seeded RSOA of 6.25, 20 and 30.7 Gbit/s in BtB and 4.2, 12.4 and 24.5 Gbit/s over 25 km NZDSF transmission are reported. To the best of our knowledge, these are the highest capacities reported for such inexpensive and colorless sources. Our experiments show that even lowest cost sources such as spectral sliced ASE sources or RSOA-FCLs can offer quite some capacity if properly operated, for example, for O-band upstream transmission. More precisely, ASE sources can provide capacities in the range of 6 to 4 Gbit/s depending on the distance. Fabry-Perot sources such as the RSOA-FCL may offer capacities in the order of 20 to 10 Gbit/s for distances up to 25 km. Finally, more costly externally seeded colorless sources may provide capacities in the range of 30 to 25 Gbit/s if operated with an optimized bit- and power loaded DMT transmission format without offset filtering.

Acknowledgments: This work was supported by the EU FP7 projects ERMES (grant agreement no. 288542) and Fox-C (grant agreement No. 318415). Sterlite Technologies Ltd. is acknowledged for providing the fiber.

Author Contributions: Simon Arega Gebrewold conceived the concept, designed and performed the experiments, carried out the signal processing and analysis and wrote the paper. Romain Bonjour contributed in the concept development, experiment and manuscript writing. Romain Brenot designed and fabricated the RSOAs. David Hillerkuss organized the experiment and wrote the paper. Juerg Leuthold conceived and developed the concept, designed the experiment, wrote the paper and supervised the work.

Conflicts of Interest: The authors declare no conflict of interest.

References

1. Effenberger, F.J.; Mukai, H.; Park, S.; Pfeiffer, T. Next-generation PON-part II: Candidate systems for next-generation PON. *IEEE Commun. Mag.* **2009**, *47*, 50–57. [CrossRef]
2. Vetter, P. Next generation optical access technologies. In Proceedings of the European Conference and Exhibition on Optical Communication, Amsterdam, The Netherlands, 16–20 September 2012; Tu.3.G.1.
3. Pfeiffer, T. Next Generation Mobile Fronthaul and Midhaul Architectures [Invited]. *J. Opt. Commun. Netw.* **2015**, *7*, 38–45. [CrossRef]
4. Saliou, F.; Gael, S.; Chanclou, P.; Pizzinat, A.; Lin, H.; Zhou, E.; Xu, Z. WDM PONs Based on Colorless Technology. *Opt. Fiber Technol.* **2015**, *26*, 126–134. [CrossRef]
5. Woodward, S.L.; Iannone, P.P.; Reichmann, K.C.; Frigo, N.J. A spectrally Sliced PON Employing Fabry-Perot lasers. *IEEE Photonics Technol. Lett.* **1998**, *10*, 1337–1339. [CrossRef]
6. Frigo, N.J. Recent Progress in Optical Access Networks. In Proceedings of the Optical Fiber Communications, San Jose, CA, USA, 25 February–1 March 1996; pp. 142–143.

7. Payoux, F.; Chanclou, P.; Genay, N. WDM-PON with Colorless ONUs. In Proceedings of the Optical Fiber Communication Conference and Exposition and the National Fiber Optic Engineers Conference, Anaheim, CA, USA, 25 March 2007.

8. Horiuchi, Y. Economical Solutions of the WDM-PON System. In Proceedings of the Optical Fiber Communication Conference, Los Angeles, CA, USA, 4–8 March 2012; pp. 1–3.

9. Healey, P.; Townsend, P.; Ford, C.; Johnston, L.; Townley, P.; Lealman, I.; Rivers, L.; Perrin, S.; Moore, R. Spectral Slicing WDM-PON Using Wavelength-Seeded Reflective SOAs. *Electron. Lett.* **2001**, *37*, 1181–1182. [CrossRef]

10. Han, K.H.; Son, E.S.; Choi, H.Y.; Lim, K.W.; Chung, Y.C. Bidirectional WDM PON using Light-Emitting Diodes Spectrum-Sliced with Cyclic Arrayed-Waveguide Grating. *IEEE Photonics Technol. Lett.* **2004**, *16*, 2380–2382. [CrossRef]

11. Arellano, C.; Polo, V.; Bock, C.; Prat, J. Bidirectional Single Fiber Transmission Based on a RSOA ONU for FTTH Using FSK-IM Modulation Formats. In Proceedings of the Optical Fiber Communication Conference, Anaheim, CA, USA, 6–11 March 2005; JWA46.

12. Buset, J.M.; El-Sahn, Z.A.; Plant, D.V. Experimental Demonstration of a 10 Gb/s RSOA-based 16-QAM Subcarrier Multiplexed WDM PON. *Opt. Express* **2014**, *22*, 1–8. [CrossRef] [PubMed]

13. Wei, J.L.; Hugues-Salas, E.; Giddings, R.P.; Jin, X.Q.; Zheng, X.; Mansoor, S.; Tang, J.M. Wavelength Reused Bidirectional Transmission of Adaptively Modulated Optical OFDM Signals in WDM-PONs Incorporating SOA and RSOA Intensity Modulators. *Opt. Express* **2010**, *18*, 9791–9808. [CrossRef] [PubMed]

14. Presi, M.; Cossu, G.; Chiuchiarelli, A.; Bottoni, F.; Corsini, R.; Choudhury, P.; Giorgi, L.; Ciaramella, E. 25 Gb/s Operation of 1-GHz Bandwidth R-SOA by using DMT and Optical Equalization. In Proceedings of the Optical Fiber Communication Conference and Exposition and the National Fiber Optic Engineers Conference, Anaheim, CA, USA, 17–21 March 2013; OW1A.7.

15. De Valicourt, G.; Maké, D.; Landreau, J.; Lamponi, M.; Duan, G.H.; Chanclou, P.; Brenot, R. High Gain (30 dB) and High Saturation Power (11 dBm) RSOA Devices as Colorless ONU Sources in Long-Reach Hybrid WDM/TDM-PON Architecture. *IEEE Photonics Technol. Lett.* **2010**, *22*, 191–193. [CrossRef]

16. Lin, G.R.; Liao, Y.S.; Chi, Y.C.; Kua, H.C.; Lin, G.C.; Wang, H.L.; Chen, Y.J. Long-Cavity Fabrey-Perot Laser Amplifier Transmitter with Enhanced Injection-Locking Bandwidth for WDM-PON Application. *J. Lightwave Technol.* **2010**, *28*, 2925–2932. [CrossRef]

17. Gebrewold, S.A.; Bonjour, R.; Barbet, S.; Maho, A.; Brenot, R.; Chanclou, P.; Brunero, M.; Marazzi, L.; Parolari, P.; Totovic, A.; et al. Self-Seeded RSOA-Fiber Cavity Lasers vs. ASE Spectrum-Sliced or Externally Seeded Transmitters—A Comparative Study. *Appl. Sci.* **2015**, *5*, 1922–1941. [CrossRef]

18. Sung-Bum, P.; Dae Kwang, J.; Dong, J.S.; Hong, S.S.; Yun, I.K.; Jeong, S.L.; Yun, K.O.; Yun, J.O. Colorless Operation of WDM-PON Employing Uncooled Spectrum-Sliced RSOA. *IEEE Photonics Technol. Lett.* **2007**, *19*, 248–250.

19. Henning, L.F.; Monteiro, P.; Almeida, A.D.; Pohl, P. Comparison of LED and RSOA performance in WDM-PONs. In Proceedings of the 21st International Conference on Telecommunications, Lisbon, Portugal, 4–7 May 2014; pp. 124–128.

20. Wong, E.; Lee, K.L.; Anderson, T.B. Directly Modulated Self-Seeding Reflective Semiconductor Optical Amplifiers as Colorless Transmitters in Wavelength Division Multiplexed Passive Optical Networks. *J. Lightwave Technol.* **2007**, *25*, 67–74. [CrossRef]

21. Marazzi, L.; Parolari, P.; Brenot, R.; De Valicourt, G.; Martinelli, M. Network-embedded self-tuning cavity for WDM-PON transmitter. *Opt. Express* **2012**, *20*, 3781–3786. [CrossRef] [PubMed]

22. Maho, A.; Gael, S.; Barbet, S.; Francois, L.; Saliou, F.; Chanclou, P.; Parolari, P.; Marazzi, L.; Brunero, M.; Martinelli, M.; et al. Demystification of the Self-Seeded WDM Access. *J. Lightwave Technol.* **2015**, *34*, 776–782. [CrossRef]

23. Le, S.D.; Lebreton, A.; Saliou, F.; Deniel, Q.; Charbonier, B.; Chanclou, P. Up to 60 km Bidirectional Transmission of a 16 Channels × 10 Gb/s FDM-WDM PON Based on Self-Seeded Reflective Semiconductor Optical Amplifiers. In Proceedings of the Optical Fiber Communications Conference and Exhibition (OFC), San Francisco, CA, USA, 9–13 March 2014; Th3G.8.

24. Shim, H.; Kim, H.; Chung, Y.C. 20-Gb/s Operation of RSOA using Polar Return-to-Zero 4-PAM Modulation Format and Direct Detection. In Proceedings of the Optical Fiber Communication Conference, Los Angeles, CA, USA, 22–26 March 2015; W1J.2.

25. Cossu, G.; Bottoni, F.; Corsini, R.; Presi, M.; Ciaramella, E. 40 Gb/s Single R-SOA Transmission by Optical Equalization and Adaptive OFDM. *IEEE Photonics Technol. Lett.* **2013**, *25*, 2119–2122. [CrossRef]
26. De Valicourt, G.; Make, D.; Fortin, C.; Enard, A.; Van-Dijk, F.; Brenot, R. 10 Gbit/s Modulation of RSOA without any Electronic Processing. In Proceedings of the Optical Fiber Communication Conference/National Fiber Optic Engineers Conference, Los Angeles, CA, USA, 6–10 March 2011; OThT2.
27. Cho, K.Y.; Takushima, Y.; Chung, Y.C. 10-Gb/s Operation of RSOA for WDM PON. *IEEE Photonics Technol. Lett.* **2008**, *20*, 1533–1535. [CrossRef]
28. Hoon, K. 10-Gb/s Operation of RSOA Using a Delay Interferometer. *IEEE Photonics Technol. Lett.* **2010**, *22*, 1379–1381.
29. Torrientes, D.; Chanclou, P.; Laurent, F.; Tsyier, S.; Chang, Y.F.; Charbonnier, B.; Raharimanitra, F. RSOA-Based 10.3 Gbit/s WDM-PON with Pre-Amplification and Electronic Equalization. In Proceedings of the Optical Fiber Communication Conference, Los Angeles, CA, USA, 21–25 March 2010; JThA28.
30. Papagiannakis, I.; Omella, M.; Klonidis, D.; Birbas, A.N.; Kikidis, J.; Tomkos, I.; Prat, J. Investigation of 10-Gb/s RSOA-Based Upstream Transmission in WDM-PONs Utilizing Optical Filtering and Electronic Equalization. *IEEE Photonics Technol. Lett.* **2008**, *20*, 2168–2170. [CrossRef]
31. Cho, K.Y.; Chung, Y.C. 10-Gb/s Operation of RSOA for WDM PON Using Return-to-Zero Modulation Format. In Proceedings of the Optical Fiber Communication Conference and Exposition (OFC/NFOEC), Los Angeles, CA, USA, 4–8 March 2012; OTh1F.2.
32. Hyun Kyu, S.; Hoon, K.; Yun, C.C. 20-Gb/s Polar RZ 4-PAM Transmission over 20-km SSMF Using RSOA and Direct Detection. *IEEE Photonics Technol. Lett.* **2015**, *27*, 1116–1119.
33. Zhang, Q.W.; Hugues-Salas, E.; Ling, Y.; Zhang, H.B.; Giddings, R.P.; Zhang, J.J.; Wang, M.; Tang, J.M. Record-high and robust 17.125 Gb/s gross-rateover 25 km SSMF transmissions of real-time dual-band optical OFDM signals directly modulated by 1 GHz RSOAs. *Opt. Express* **2014**, *22*, 6339–6348. [CrossRef] [PubMed]
34. Xie, C.; Dong, P.; Randel, S.; Winzer, P.; Spiga, S.; Koegel, B.; Neumeyr, C.; Amann, M.C. Single-VCSEL 100-Gb/s Short-Reach System Using Discrete Multi-Tone Modulation and Direct Detection. In Proceedings of the Optical Fiber Communications Conference and Exhibition (OFC), Los Angeles, CA, USA, 22–26 March 2015; pp. 1–3.
35. Takahara, T.; Tanaka, T.; Nishihara, M.; Kai, Y.; Li, L.; Tao, Z. Discrete Multi-Tone for 100 Gb/s Optical Access Networks. In Proceedings of the Optical Fiber Communication Conference, San Francisco, CA, USA, 9–13 March 2014; M2l.1.
36. Nadal, L.; Moreolo, M.S.; Fabrega, J.M.; Dochhan, A.; Griesser, H.; Eiselt, M.; Elbers, J.P. DMT Modulation With Adaptive Loading for High Bit Rate Transmission Over Directly Detected Optical Channels. *J. Lightwave Technol.* **2014**, *32*, 4143–4153. [CrossRef]
37. Gebrewold, S.A.; Brenot, R.; Bonjour, R.; Josten, A.; Baeuerle, B.; Hillerkuss, D.; Hafner, C.; Leuthold, J. Colorless Low-Cost RSOA Based Transmitters Optimized for Highest Capacity Through Bit- and Power-Loaded DMT. In Proceedings of the Optical Fiber Communication Conference, Anaheim, CA, USA, 20–22 March 2016; Tu2C.4.
38. Chow, P.S.; Cioffi, J.M.; Bingham, J.A. A practical Discrete Multitone Transceiver Loading Algorithm for Data Transmission over Spectrally Shaped Channels. *IEEE Trans. Commun.* **1995**, *43*, 773–775. [CrossRef]
39. Pollet, T.; Peeters, M. Synchronization with DMT modulation. *IEEE Commun. Mag.* **1999**, *37*, 80–86. [CrossRef]
40. Nakagawa, T.; Kobayashi, T.; Ishihara, K.; Miyamoto, Y. Non-Data-Aided Wide-Range Frequency Offset Estimator for QAM Optical Coherent Receivers. In Proceedings of the Optical Fiber Communication Conference/National Fiber Optic Engineers Conference, Los Angeles, CA, USA, 6–10 March 2011; OMJ1.
41. Schmogrow, R.; Nebendahl, B.; Winter, M.; Josten, A.; Hillerkuss, D.; Koenig, S.; Meyer, J.; Dreschmann, M.; Huebner, M.; Koos, C.; et al. EVM as a Performance Measure for Advanced Modulation Formats. *IEEE Photonics Technol. Lett.* **2012**, *24*, 61–63. [CrossRef]
42. Martinelli, M.; Marazzi, L.; Parolari, P.; Brunero, M.; Gavioli, G. Polarization in Retracing Circuits for WDM-PON. *IEEE Photonics Technol. Lett.* **2012**, *24*, 1191–1193. [CrossRef]
43. Marazzi, L.; Parolari, P.; Boletti, A.; Gatto, A.; Martinelli, M.; Brrnot, R. Highly-nonlinear RSOA RIN compression. In Proceedings of the 19th European Conference on Networks and Optical Communications, Milano, Italy, 4–6 June 2014; pp. 115–119.

44. Kani, J.; Kawata, H.; Iwatsuki, K.; Ohki, A.; Sugo, M. Design and Demonstration of Gigabit Spectrum-Sliced WDM Systems Employing Directly Modulated Super Luminescent Diodes. In Proceedings of the Optical Fiber Communication Conference and Exposition and the National Fiber Optic Engineers Conference, Anaheim, CA, USA, 6–11 March 2005; JWA49.

45. Pendock, G.J.; Sampson, D.D. Transmission performance of high bit rate spectrum-sliced WDM systems. *J. Lightwave Technol.* **1996**, *14*, 2141–2148. [CrossRef]

46. Wei, J.L.; Hamie, A.; Gidding, R.P.; Hugues-Salas, E.; Zheng, X.; Mansoor, S.; Tang, J.M. Adaptively Modulated Optical OFDM Modems Utilizing RSOAs as Intensity Modulators in IMDD SMF Transmission Systems. *Opt. Express* **2010**, *18*, 8556–8573. [CrossRef] [PubMed]

47. Marazzi, L.; Parolari, P.; Brunero, M.; Gatto, A.; Martinelli, M.; Brenot, R.; Barbet, S.; Galli, P.; Gavioli, G. Up to 10.7-Gb/s High-PDG RSOA-Based Colorless Transmitter for WDM Networks. *IEEE Photonics Technol. Lett.* **2013**, *25*, 637–640. [CrossRef]

48. Petermann, K. Noise Characterstics of Solitary Laser Diodes. In *Laser Diode Modulation and Noise*; Okoshi, T., Ed.; Kluwer Academic Publishers: Dordrecht, The Netherlands, 1998; pp. 152–208.

49. Ma, Y.; Xu, Z.; Lin, H.; Zhou, M.; Wang, H.; Zhang, C.; Yu, J. Demonstration of Digital Fronthaul Over Self-Seeded WDM-PON in Commercial LTE Environment. *Opt. Express* **2015**, *23*, 11927–11935. [CrossRef] [PubMed]

50. Gebrewold, S.A.; Marazzi, L.; Parolari, P.; Brenot, R.; Duill, P.O.S.; Bonjour, R.; Hillerkuss, D.; Hafner, C.; Leuthold, J. Reflective-SOA Fiber Cavity Laser as Directly Modulated WDM-PON Colorless Transmitter. *IEEE J. Sel. Top. Quantum Electron.* **2014**, *20*, 1–9. [CrossRef]

51. Gebrewold, S.A.; Marazzi, L.; Parolari, P.; Brunero, M.; Brenot, R.; Hillerkuss, D.; Hafner, C.; Leuthold, J. Colorless Self-Seeded Fiber Cavity Laser Transmitter for WDM-PON. In Proceedings of the CLEO: Science and Innovations, San Jose, CA, USA, 8–13 June 2014; STu1J.4.

52. André, N.S.; Habel, K.; Louchet, H.; Richter, A. Adaptive Nonlinear Volterra Equalizer for Mitigation of Chirp-Induced Distortions in Cost Effective IMDD OFDM Systems. *Opt. Express* **2013**, *21*, 26527–26532. [CrossRef] [PubMed]

53. Simon, G.; Saliou, F.; Chanclou, P.; Deniel, Q.; Erasme, D.; Brenot, R. 70 km external cavity DWDM sources based on O-band Self Seeded RSOAs for transmissions at 2.5 Gbit/s. In Proceedings of the Optical Fiber Communication Conference, San Francisco, CA, USA, 9–13 March 2014; W3G.5.

54. Zhang, L.M.; Kschischang, F.R. Staircase Codes with 6% to 33% Overhead. *J. Lightwave Technol.* **2014**, *32*, 1999–2002. [CrossRef]

55. Bonk, R.; Brenot, R.; Meuer, C.; Vallaitis, T.; Tussupov, A.; Rode, J.C.; Sygletos, S.; Vorreau, P.; Lelarge, F.; Duan, G.H.; et al. 1.3/1.5 µm QD-SOAs for WDM/TDM GPON with Extended Reach and Large Upstream/Downstream Dynamic Range. In Proceedings of the Optical Fiber Communication—Incudes post deadline papers, San Diego, CA, USA, 22–26 March 2009; OWQ1.

applied
sciences

MDPI

Article

Theoretical Analysis of Directly Modulated Reflective Semiconductor Optical Amplifier Performance Enhancement by Microring Resonator-Based Notch Filtering

Zoe V. Rizou * and Kyriakos E. Zoiros

Lightwave Communications Research Group, Laboratory of Telecommunications Systems,
Department of Electrical and Computer Engineering, Democritus University of Thrace, 67 100 Xanthi,
Greece; kzoiros@ee.duth.gr
* Correspondence: zrizou@ee.duth.gr; Tel.: +30-25410-79-975

Received: 30 December 2017; Accepted: 28 January 2018; Published: 1 February 2018

Abstract: We demonstrate the feasibility of using a single microring resonator (MRR) as optical notch filter for enabling the direct modulation of a reflective semiconductor optical amplifier (RSOA) at more than tripled data rate than possible with the RSOA alone. We conduct a thorough simulation analysis to investigate and assess the impact of critical operating parameters on defined performance metrics, and we specify how the former must be selected so that the latter can become acceptable. By using an MRR of appropriate radius and detuning, the RSOA modulation bandwidth, which we explicitly quantify, can be extended to overcome the RSOA pattern-dependent performance limitations. Thus, the MRR makes the RSOA-encoded signal exhibit improved characteristics that can be exploited in practical RSOA direct modulation applications.

Keywords: direct modulation; microring resonator; optical notch filter; reflective semiconductor optical amplifier

1. Introduction

Reflective semiconductor amplifiers (RSOAs) are key modules for the realization of next-generation broadband access applications, such as colorless passive optical access networks [1], radio over fiber [2], slow and fast light [3], data erasing and remodulation [4], information packet power equalization [5] and fiber-optical cable television [6], which rely on the manipulation of data flowing in opposite directions. In effect, a single RSOA can simultaneously receive and amplify signals in the downstream while remodulating them with end-user information in the upstream communication link. This multifunctional capability is inherent to RSOAs, which owing to their internal structure use the same device facet both for signal input injection and output extraction. Thus, RSOAs avoid the need of having extra fiber paths, fiber components and optoelectronic elements to support bidirectional transmission, which helps reduce the overall system cost and complexity. However, as a physical byproduct of semiconductor optical amplifiers' finite carrier lifetime [7], the modulation bandwidth, which scales inversely against this parameter [8], of these devices, including RSOAs, is limited to a few GHz. For this reason, an RSOA can be directly modulated by electrical data, which are encoded on a seeding continuous wave (CW) light, only up to speeds which are not sufficient for satisfying the increasing bandwidth needs of the target applications. In the effort to overcome this fundamental limitation towards enabling full-duplex and symmetrical RSOA-based data transfer, several promising options are available, which include electronic equalization [9], advanced modulation formats [10,11], specially designed RSOA package and driving circuitry [12,13], and optical notch filtering [14]. In particular, the latter has been widely exploited because of the simple configuration, cost-affordable implementation using off-the-shelf components and passive nature of the underlying signal-equalization mechanism. Furthermore, it is not concerned with challenging issues of the three other alternatives, such as the sensitivity to fiber chromatic dispersion and the requirement of high-speed sophisticated electronics at the transmitter and receiver side, in the first; the inevitable complex signal generation and detection in the second; and the involvement of elaborate fabrication and integration processes in the third. In fact, the two first RSOA bandwidth-extension techniques have even been combined with optical equalization and benefited from its efficiency to further improve their performance [11,15,16]. Thus, different schemes that act as optical notch filters on RSOA-encoded outputs have been reported, such as the delay interferometer (DI) [17,18], the fiber Bragg grating (FBG) [19] and the arrayed waveguide grating (AWG) [20]. In the same technological category also fall schemes that have been applied on conventional directly modulated SOAs, such as the birefringent fiber loop (BFL), either alone [21], in cascade with another BFL [22] or assisted by an optical bandpass filter [23]; and the microring resonator (MRR) [24]. The MRR is a special form of powerful waveguides [25,26], and exhibits distinctive characteristics over the aforementioned types of employed filters. These include the structural simplicity, the ultra-compact size–which makes it amenable to integration with microelectronic fabrication processes–enhanced wavelength selectivity, fine-tuning capability and availability of different material systems with the potential of copackaging with the SOA in the same hybrid platform. These attributes render the MRR a mostly suitable solution for increasing the operating speed of directly modulated RSOAs as well. In this paper, we investigate and demonstrate, by means of theoretical simulation and analysis, the MRR potential for overcoming the RSOA limited modulation bandwidth and thus for being extended beyond classical filter-oriented applications. For this purpose, we apply the model proposed in [27] to theoretically describe the operation of an RSOA when directly modulated by an electrical data pattern, a task that we initiated within the frame of [28]. Unlike the usually followed modeling approach, which involves solving a set of coupled partial nonlinear differential equations with boundary conditions [29,30], and hence is computationally demanding, this reduced model has been formulated on the basis of valid assumptions and prior framework [31] in such way that it allows one to derive the encoded signal at the RSOA output from the solution of a single standard differential equation in the time domain. This fact greatly simplifies the computational complexity while still providing realistic and accurate results. In this manner, we have been able to explore the capability of the MRR to enhance the data rate at which the RSOA can be directly modulated with acceptable performance. We have thus confirmed that if the MRR critical parameters are selected as specified in this work, it is indeed possible to extend by more than three times the repetition rate of the encoded signal, with the latter exhibiting improved characteristics.

2. RSOA Direct Modulation Assisted by MRR-Based Notch Filtering

2.1. Configuration

The basic configuration of a directly modulated RSOA is shown in Figure 1. A CW signal of constant power, P_{CW}, is inserted in the RSOA. Concurrently, a radio frequency (RF) signal, which comprises non-return-to-zero (NRZ) data pulses, is superimposed to the RSOA DC bias current, I_{bias}, and induces, via the RSOA electrical impedance, a peak current offset, $\pm I_m$. Since the RSOA rear edge is highly reflective, the CW signal undertakes a double pass inside the active medium of length L, and when it exits from the RSOA front facet, which is highly antireflective, it has perceived the gain variations due to the modulated current. Normally, the CW signal at the RSOA output should bear the exact digital information of the applied electrical excitation in optical form. However, as the rate of direct modulation gets faster, this may not happen, as due to the RSOA limited modulation bandwidth, the performance of the directly modulated RSOA is progressively deteriorated by pattern effects and eventually becomes poor. Still, it is possible to enhance the RSOA direct modulation capability by means of optical notch filtering. For this purpose, the RSOA is connected to an MRR, which, as shown in Figure 1, is a waveguide shaped into a ring structure of radius R coupled to a bus waveguide with field transmission coefficient r.

By properly tailoring the MRR spectral response, the MRR can efficiently act as frequency discriminator on the encoded pulses and improve their quality. This is done by adjusting the wavelength separation between adjacent peaks or notches, or free spectral range (FSR), which is inversely proportional to the MRR radius, and wavelength offset, that is, detuning, $\Delta\lambda$, between the encoded signal spectral position, λ_{enc}, and the nearest transfer function (TF) transmission peak, while maximizing the MRR transfer function peak-to-notch contrast ratio (PNCR) through operating the MRR in the critical coupling regime, where the microring internal losses are equal to the coupling losses. In this manner, the MRR compensates for the inherently limited bandwidth of the RSOA, and mitigates the pattern-dependent impairments induced on the encoded pulses.

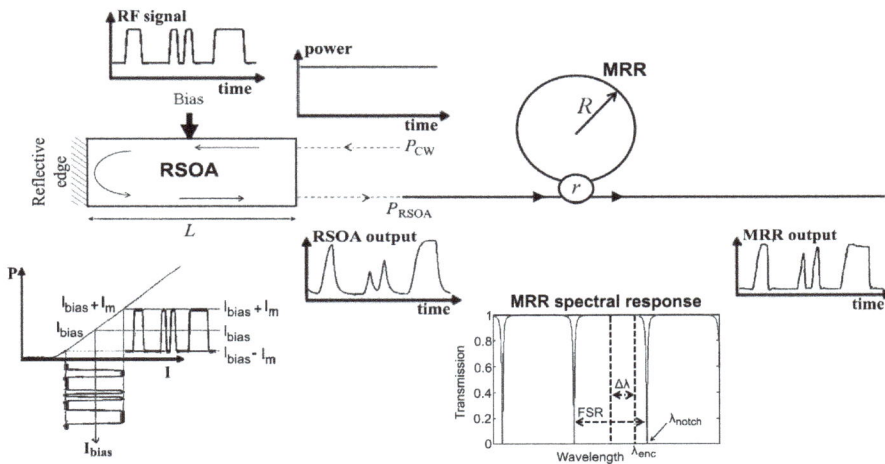

Figure 1. Conceptual view of reflective semiconductor amplifier (RSOA) direct modulation assisted by serially connected microring resonator (MRR). FSR: free spectral range, RF: radio frequency.

2.2. Modeling

The performance of the setup in Figure 1 is characterized by the power that emerges at the output of the directly modulated RSOA, $P_{RSOA}(t)$, and of the cascaded MRR-based notch filter, $P_{MRR}(t)$. This means that in order to simulate the operation of the scheme, it is necessary to know and calculate at each one of these points the lightwave-encoded signal electric fields, since they are normalized so that their squared modulus represents power, that is, $P_{RSOA,MRR}(t) = |E_{RSOA,MRR}(t)|^2$ [32].

For the RSOA case

$$E_{RSOA}(t) = \sqrt{P_{CW}} \exp\left[(1 - j\alpha_{LEF})h(t - 2Ln_g/c)\right], \tag{1}$$

where α_{LEF} is the RSOA linewidth enhancement factor, n_g is the group refractive index of the semiconductor material, c is the speed of light in vacuum, and the RSOA gain response integrated over its length is denoted by $h(t)$, which is shifted in time by twice the RSOA one-way transit time so as to account for the double pass taken by the lightwave signal within the RSOA optical cavity. Now, function $h(t)$ obeys the following one-dimensional differential equation [27]

$$\frac{dh(t)}{dt} = -\frac{h(t) - \Gamma a N_0 L\left[\dfrac{I(t)}{I_0} - 1\right]}{T_{car}} - \frac{\exp[2h(t)] - 1}{E_{sat}} P_{CW}, \tag{2}$$

where Γ is the RSOA confinement factor, a is the RSOA differential gain, N_0 is the RSOA carrier density at transparency, T_{car} is the RSOA carrier lifetime, $I_0 = qALN_0/T_{car}$ is the RSOA current at transparency (where q is the electron charge and A is the area of the semiconductor active region), and E_{sat} is the RSOA saturation energy. Equation (2) has been obtained after making the following valid assumptions and simplifications: (a) The RSOA internal losses are neglected, as they can be well compensated by the RSOA CW gain, G_{CW}, through the adjustment of the RSOA bias current, since these two parameters are directly related [27]; (b) the RSOA reflectivity is considered perfect, that is, 100%, as in practice this is a basic feature of the specific devices that favors their use [33], in particular as intensity modulators [10]; and (c) the RSOA round-trip propagation time, which is determined by the RSOA active region length, is smaller than the pulse repetition interval of the applied excitation. This condition is satisfied for the direct modulation rates pursued in RSOAs being less than 1 mm long [34].

The profile of each pulse inside the $N = 2^7 - 1$ bit-long NRZ pseudorandom binary sequence (PRBS) of the injection current is described in compact form [35,36] as the sum of a DC term and a modulation term

$$I(t) = \overbrace{(I_{bias} - I_m)}^{\text{DC term}} + \overbrace{2I_m \left\{ H(t) \left[1 - e^{-t^2/t_r^2} \right] - H(t-T) \left[1 - e^{-(t-T)^2/t_r^2} \right] \right\}}^{\Delta I(t) \equiv \text{ modulation term}}, \qquad (3)$$

where $H(t)$ is Heaviside step function, I_{bias} and I_m are the RSOA bias current and induced RF modulation current, respectively–which are a small fraction of the current at transparency–and t_r is the pulse rise time, which occupies a small portion of the pulse repetition period, T.

For the MRR case, obtaining the electric field at its output involves the following steps. First, the RSOA temporal response is extracted by numerically solving for $h(t)$ from (2). This is done by following the numerical method formulated in [37] but adapted to account for a temporal-dependent total injection current [38]. More specifically, for each information bit of duration T carried by the modulation current, (2) is solved in a step-wise manner by approximating the time derivative of $h(t)$ by a finite difference and applying the appropriate initial conditions. For this purpose, the continuous time variable t is replaced by discrete points $t_{p,i} = (p-1)T + i\Delta t$, for $p = 1, 2, \ldots, N$ and $i = 1, 2, \ldots, k$, where p denotes the p-th pulse inside the N-long PRBS of the injection current, Δt is the finite temporal increment and $k = T/\Delta t$ is the integer number of calculated samples of function $h(t)$ in each bit period. To ensure high enough temporal resolution and at the same time enhanced computational efficiency so as to correctly capture the RSOA gain dynamics expressed by $h(t)$ with affordable hardware and software resources, we choose $\Delta t = 1$ ps, which means that nearly 100 samples are taken at all data rates considered in this paper. Then, the knowledge of $h(t)$ and of its derivative at a given pulse instant "i", $h_{p,i}$ and $dh_{p,i}/dt$, respectively, allows one to calculate the value, $h(t + \Delta t)$, at the next discrete moment, $h_{p,i+1}$, according to Euler's numerical method. This method is suitable for studying the response of semiconductor active waveguide devices, such as the RSOA, to an electrical excitation of piecewise varying nature, such as the injection current NRZ pulses, as it converges rather fast while producing reasonable results [39]. The initial conditions required for running this process are (a) $h_{1,1} = \bar{g}_0 L$, where $\bar{g}_0 = \Gamma a N_0 \{ [(I_{bias} - I_m)/I_0] - 1 \}$ is the RSOA steady-state gain coefficient. This condition is a direct byproduct of the fact that the first temporal segment in the leading edge of the very first information pulse, which is assumed to be a mark and is supplied to the RSOA through the injection current pulse stream, experiences an unsaturated RSOA gain, since the carriers of the latter have not had the time to be modified yet; (b) $h_{p,1} = h_{p-1,k}$, to account for the fact that the gain available to the first temporal segment in the leading edge of every next pulse is that left by the last temporal segment of the immediately preceding pulse. This condition is properly adapted and reduced to the case of consecutive NRZ pulses carrying the same binary information, that is, strings of marks or spaces. This process is carried out iteratively until all $N \cdot k$-th values of $h(t)$ are obtained, which can be then stored in a matrix and retrieved for further calculations. The solution for $h(t)$ is then substituted in (1) to find the electric field of the encoded signal at the RSOA output. This function is transferred then into the frequency domain and is convolved with the MRR spectral response. The latter has the compact mathematical expression given below [40],

$$T_{MRR}(\lambda) = \frac{r - l \exp\left[jn_{eff}4\pi^2 R\left((\lambda - \Delta\lambda)/\lambda^2\right)\right]}{1 - rl \exp\left[jn_{eff}4\pi^2 R\left((\lambda - \Delta\lambda)/\lambda^2\right)\right]}, \tag{4}$$

where n_{eff} is the waveguide effective refractive index while the field transmission coefficient, r, and amplitude attenuation factor, l, are equal and tend to unity so as to ensure that the MRR is operating in the critical coupling regime [41] where it can efficiently act as notch filter [42]. Then, the convolution product is converted back into the time domain. This procedure can be put in mathematical form as $E_{MRR}(t) = \mathcal{F}^{-1}\{\mathcal{F}[E_{SOA}(t)]T_{MRR}(\lambda)\}$, where operators, $\mathcal{F}\{\cdot\}$ and $\mathcal{F}^{-1}\{\cdot\}$ denote fast Fourier transform (FFT) and inverse FFT, respectively, which are both available and executed in Matlab software. Therefore, $P_{MRR}(t) = |E_{MRR}(t)|^2$. In this manner, the maximum and average values of the peak power of the marks and spaces within the encoded PRBS can be found so that they can be used to calculate the performance metrics employed in the following section.

The operation of the directly modulated RSOA, either alone or assisted by the MRR, is also characterized by the respective modulation responses, which are derived through the process of small-signal analysis [43,44]. Starting with the RSOA, we have

$$T_{RSOA} \overset{\text{①}}{=} \frac{p_{RSOA}(\Omega_m)}{p_{RSOA}(0)} \overset{\text{②}}{=} \frac{\Delta h(\Omega_m)}{\Delta h(0)} \overset{\text{③}}{=} \frac{\Delta g(\Omega_m)D(0)}{\Delta g(0)D(\Omega_m)}, \tag{5}$$

where the expressions that follow after steps ①, ② and ③ have been derived by combining relevant information from [18,27] and [45] (c.f. Appendix). More specifically, in ①, $p_{RSOA}(\Omega_m)$ is the small-signal power, which is produced at the RSOA output due to the modulation of its current at a given frequency, $f_m = \Omega_m/2\pi$, and normalized over its unmodulated counterpart [46]; in ②, $\Delta h(\Omega_m)$ is the concomitant small-signal deviation from steady state of the RSOA-integrated gain given by [27] $\Delta h(\Omega_m) = [L/D(\Omega_m)]\Delta g(\Omega_m)$, where $\Delta g(\Omega_m) = \Gamma a N_0 \Delta I(\Omega_m)/I_0$ is the corresponding gain coefficient perturbation incurred by the modulation current of complex envelope $\Delta I(\Omega_m)$; and in ③, $D(\Omega_m) = 1 - j\Omega_m T_{car} + W\left[\frac{2T_{car}P_{CW}}{E_{sat}}\exp\left(2\bar{g}_0 L + \frac{2T_{car}P_{CW}}{E_{sat}}\right)\right]$ [27], where $W[\cdot]$ is Lambert's 'W' function [47]. Function $D(\Omega_m)$ links in a compact manner the perturbation of the RSOA-integrated gain to the current modulation from which it has been incurred. Thus, when the RSOA is directly modulated by a sinusoidal electrical excitation, it allows, through first-order approximations, one to analytically express the 3 dB angular frequency of the RSOA modulation response as [27] $\Omega_{3dB} = \frac{\sqrt{3}}{T_{car}}\left\{1 + W\left[\frac{2T_{car}P_{CW}}{E_{sat}}\exp\left(2\bar{g}_0 L + \frac{2T_{car}P_{CW}}{E_{sat}}\right)\right]\right\}$. This formula provides an estimate of the maximum RSOA modulation frequency and hence of its direct modulation capability at a given data rate [28]. In general, however, the modulation current may be in digital form and have any shape. Then, the 3-dB modulation bandwidth can be derived from (5) by taking the Fourier transform of the modulation current in the time domain, which in our case is given by the second term in the right-hand side of (3). By using Fourier transform formulas and properties of the functions involved therein, and after tight algebraic manipulations, in the course of which the complementary error function and the imaginary error function come into being, we find the explicit expression

$$\Delta I(\Omega_m) = \mathcal{F}[\Delta I(t)] =$$
$$= 2I_m\left\{\frac{1}{j\Omega_m}\left[1 - \exp(-j\Omega_m T)\right] + \left[\exp(-j\Omega_m T) - 1\right]\left[0.5\sqrt{\pi}t_r \exp\left(-\Omega_m^2 t_r^2/4\right) - jt_r\mathbb{F}(\Omega_m t_r/2)\right]\right\}, \tag{6}$$

where function $\mathbb{F}(x) = \exp(-x^2)\int_0^x \exp(t^2)dt$ represents Dawson's integral [48], which is available in Matlab software. Taking the squared modulus and after proper algebraic manipulations, (6) can be brought into a form that allows one to approximate its values for the boundary conditions of the modulation frequency [46]. The obtained result is analogous to the square of the product of the electrical modulation pulses' peak amplitudes and period when $\Omega_m \to 0$, while it tends to null when $\Omega_m \to \infty$. In other words, the level of the specific function in these two limits is in the opposite direction to that of the modulation frequency variation. This fact confirms that the RSOA modulation response exhibits a low-pass characteristic, which is in agreement with the same trend noticed from extensive numerical calculations [49].

For the MRR, on the other hand, the modulation response is [24]

$$T_{MRR} = A \times ((1 - j\alpha_{LEF})/2) \times \frac{A^* + \Omega_m^2 + 2\delta\Omega_m + j\mu_e^2\Omega_m}{\delta^2\left((\delta + \Omega_m)^2 + 1/\tau^2\right)} + \\ + A^* \times ((1 + j\alpha_{LEF})/2) \times \frac{A + \Omega_m^2 - 2\delta\Omega_m + j\mu_e^2\Omega_m}{\delta^2\left((\delta - \Omega_m)^2 + 1/\tau^2\right)}. \tag{7}$$

In this expression, $A = \delta^2 - j\mu_e^2\delta$ (and A^* is the complex conjugate), $\delta = 2\pi c\frac{\lambda_{notch} - \lambda_{enc}}{\lambda_{notch}\lambda_{enc}}$, where λ_{notch} is the spectral position of the MRR transfer function notch (Figure 1) that obeys the condition of MRR resonance, $2\pi Rn_{eff} = m\lambda_{notch}$, $m \in Z^*$, and lies nearest to the encoded optical signal of wavelength, λ_{enc}, in the vicinity of 1550 nm, and μ_e is the coupling strength between the bus waveguide and the ring, $\mu_e^2 = \left[(1 - r^2)c/2\pi Rn_{eff}\right] = 1/\tau$, where the last equality with the MRR $1/e$ amplitude decay time, τ, holds under critical coupling [44]. The modulation response of the MRR-assisted RSOA is then obtained by taking the absolute value of (5) and multiplying it with the modulus of (7).

Finally, an inevitable consequence of RSOA direct modulation is that the encoded pulses acquire an instantaneous frequency deviation, that is, chirp, across them [50]. The compensation of the chirp's irregular temporal variation via post-optical notch filter is converted by the latter to restored quality of encoded pulses and accordingly enhanced RSOA modulation capability. This means that quantifying this transient effect can also provide useful information for the operation and performance of the MRR-assisted directly modulated RSOA. For this purpose, the knowledge of the phase response of the RSOA and MRR is required since the chirp, $\Delta v(t)$, is by definition linked to it through the derivative in the time domain [51]. Finding the chirp for the SOA case is straightforward by isolating the output phase term from (1), differentiating it in time and readily taking the differential of the time-delayed version of the RSOA-integrated gain response from the right-hand side of (2) using the transform $t \to t - 2Ln_g/c$. For the MRR case, however, the phase of the transmitted light undergoes a steep variation around resonance [52], which compromises the direct numerical calculation of the corresponding response and accordingly of its temporal rate of change, which provides the chirp. Instead of taking the inverse tangent function of the ratio between the imaginary and real parts of the complex-valued electric field [53], we use the relevant information that is extracted when applying the signal phase-reconstruction technique in the optical domain [54,55]. This technique allows one to unambiguously recover the instantaneous frequency deviation profile of a random repetitive data signal which is inserted in a frequency discriminator whose spectral transfer function has a linear spectral amplitude variation around the detuning position. This element acts as a differentiator on the inserted signal, which is exactly what the MRR-based notch filter does in our case [52]. In this manner, the desired chirp can be directly extracted in analytical form from the knowledge of the time-domain intensity profiles of the signals that enter and exit the MRR. Thus [54],

$$-2\pi\Delta v_{MRR}(t) = -\sqrt{\frac{1}{|E_{RSOA}(t)|^2}\cdot\left[\left(\frac{|E_{MRR}(t)|}{S}\right)^2 - \left(\frac{\partial|E_{RSOA}(t)|}{\partial t}\right)^2\right]} + \delta, \tag{8}$$

where S is the slope of the linear amplitude variation around the detuning point of the MRR TF with respect to angular frequency, which can be found by differentiating (4) and applying the chain rule [56].

3. Results

We found first if and how the RSOA can be directly modulated on its own without performance degradation. For this purpose we investigated the impact of the parameters associated with the optical and electrical RSOA excitation signal on the error probability (EP) which, for acceptable performance, should not exceed the forward error correction (FEC) threshold of 3.8×10^{-3} [22,24]. The error probability is analytically extracted through the Q-factor, which is defined as

$$Q = \frac{\bar{P}_1 - \bar{P}_0}{\sigma_{1,pe} + \sigma_{0,pe}},$$ (9)

where $\bar{P}_1$, $\bar{P}_0$ are the mean and $\sigma_{1,pe}^2$, $\sigma_{0,pe}^2$ are the variances of the peak power of encoded marks ('1's) or spaces ('0's), respectively, which are numerically computed according to the details provided in the previous section, in the presence of pattern effects manifested due to the RSOA direct modulation (hence the subscript 'pe'). Because the amplitude distortions incurred by these effects on the encoded bits at the RSOA output are intense, the variances in the denominator of (9) dominate over the variances of conventional noise components, such as those associated with the RSOA amplified spontaneous emission (ASE) [30]. Moreover, given that the statistics of these distortions follow the Gaussian distribution [30], the relationship between the EP and the Q-factor reads [57]

$$EP = \frac{1}{2}\text{erfc}\left(\frac{Q}{\sqrt{2}}\right),$$ (10)

where erfc(.) is the complementary error function.

Figure 2 shows the EP versus the RSOA CW input power and bias current for three different direct modulation rates. This figure has been obtained for the RSOA structural and physical parameter values employed or derived in [28]. The values of the total injection current characteristics that have been used in deriving Figure 2a are $I_{bias} = 1.2I_0$ and $I_m = 0.1I_0$, where the current at transparency is ~75 mA and $t_r = 17\%T$. On the other hand, Figure 2b has been obtained for a CW input power of -10 dBm. The EP is acceptable for all scanned parameters' range at 2.5 Gb/s, but as we go to 3 Gb/s and beyond, it is necessary to provide more CW power and bias current. The physical explanation for this behavior of the EP is that the increase of the RSOA driving optical power and applied electrical bias accelerates the RSOA response [10] and widens its modulation bandwidth [58], so that the RSOA is allowed to handle data of higher rate. According to this fact, the continuous increase of the examined parameter values would seem natural for supporting faster RSOA direct modulation. However, there are physical limitations in this trend, which are imposed by the RSOA CW gain and the maximum amplitude difference, that is, extinction, between marks and spaces, $AD_{1/0,max}$ [24]. In fact, increasing too much the CW power forces the RSOA to enter into the deep saturation regime where it provides a reduced optical gain, while increasing too much the bias current shifts the RSOA electrical modulation to occur beyond the linear gain–current region, that is, into the hard limit where the information-driving current is made to experience an almost flat RSOA optical gain [58]. For these physical reasons, the electrical current-induced RSOA optical gain perturbation conditions become such that they degrade the encoded output quality due to the modulated signal's decreased extinction in the first case and increased clipping in the second one.

The CW gain should not fall below 10 dB so that the encoded signal, after undergoing the subsequent filtering action, can still have enough power to be used in RSOA direct modulation applications [13]. $AD_{1/0,max}$ should exceed 10 dB [13] so that the encoded marks are sufficiently distinguished from the encoded spaces.

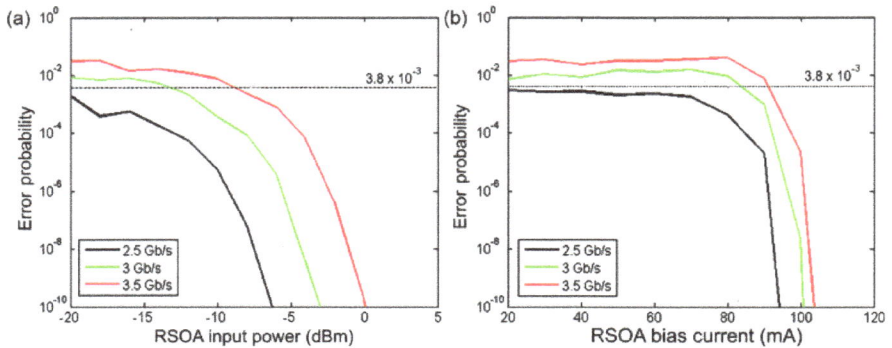

Figure 2. Error probability at RSOA output versus RSOA (**a**) input power and (**b**) bias current. The horizontal dotted line denotes the FEC limit.

Then, from Figures 3 and 4 it can be observed that these requirements can simultaneously be satisfied for some input power and bias current values only up to 3 Gb/s. Thus, an appropriate combination of these parameters is $P_{CW} = -12$ dBm and $I_{bias} = 90$ mA, for which $EP = 4.3 \times 10^{-3}$, CW gain = 12.05 dB and $AD_{1/0,max} = 10.02$ dB.

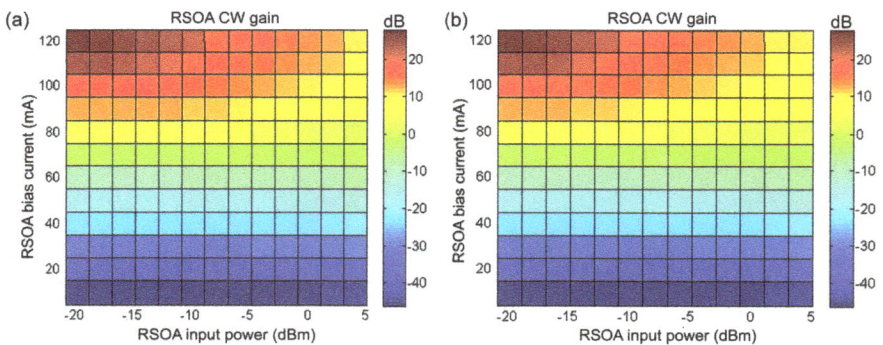

Figure 3. RSOA continuous wave (CW) gain versus RSOA input power and bias current at (**a**) 3 Gb/s and (**b**) 3.5 Gb/s.

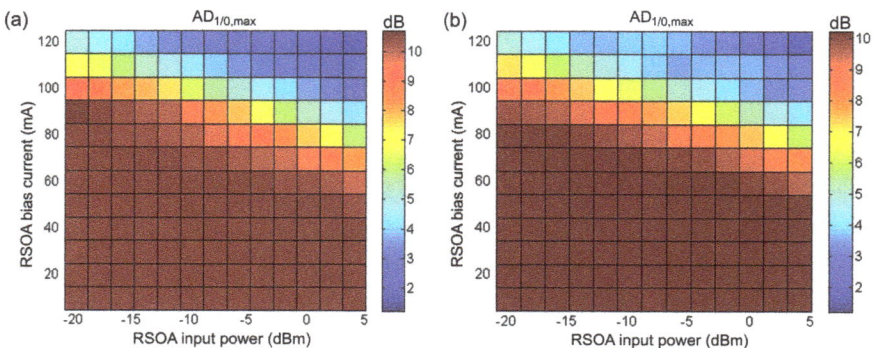

Figure 4. $AD_{1/0,max}$ versus RSOA input power and bias current at (**a**) 3 Gb/s and (**b**) 3.5 Gb/s.

The suitability of this choice is also confirmed by the resultant RSOA modulation bandwidth, which, as shown in Figure 5, is of the order of 3 GHz and hence consistent with the above findings.

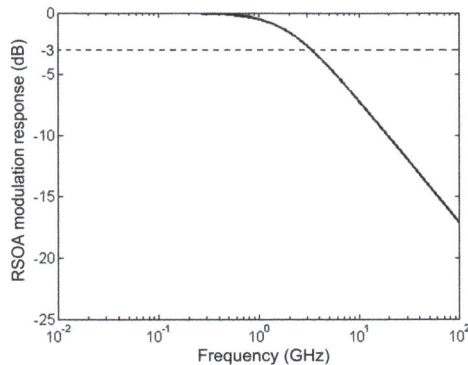

Figure 5. RSOA modulation response for input power -12 dBm and bias current 90 mA.

On the other hand, Figure 6 shows that the RSOA direct modulation capability is not affected by the injection current modulation amplitude and rise time since, despite their variation, the RSOA modulation bandwidth remains nearly the same. Physically this happens because the RSOA operating conditions set by the specified CW power and bias current are such that the negative effects of encoded signal extinction ratio degradation and amplitude clipping are appropriately balanced so that they become rather independent of the modulation current peak and rise time.

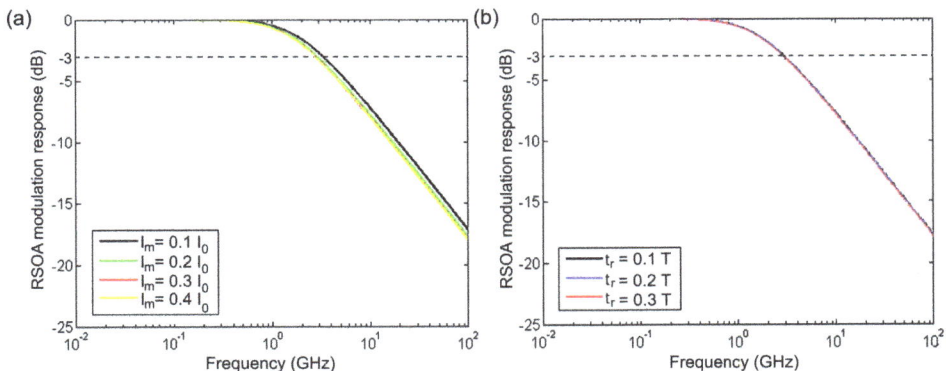

Figure 6. RSOA modulation response for different (**a**) peak modulation currents and (**b**) modulation current rise times.

Figure 7 compiles the encoded pulse waveforms, chirp and pseudo-eye diagrams (PEDs) for favorable (top) and adverse (bottom) RSOA direct modulation which occur at 3 Gb/s and 5 Gb/s, respectively. These opposing cases can be quantitatively compared not only against the EP but also against the maximum amplitude difference between marks, $AD_{1,max}$, and $AD_{1/0,max}$ [24], red chirp fluctuations (CF) [59] and PED eye opening (EO) [24]. Thus, $EP = 4.3 \times 10^{-3}$ at 3 Gb/s vs. 7.6×10^{-2} at 5 Gb/s, $AD_{1,max} = 0.22$ dB at 3 Gb/s vs. 1.03 dB at 5 Gb/s, $AD_{1/0,max} = 10.02$ dB at 3 Gb/s vs. 8.1 dB at 5 Gb/s, $CF = 2\%$ at 3 Gb/s vs. 15% at 5 Gb/s and $EO = 87\%$ at 3 Gb/s vs. 72% at 5 Gb/s.

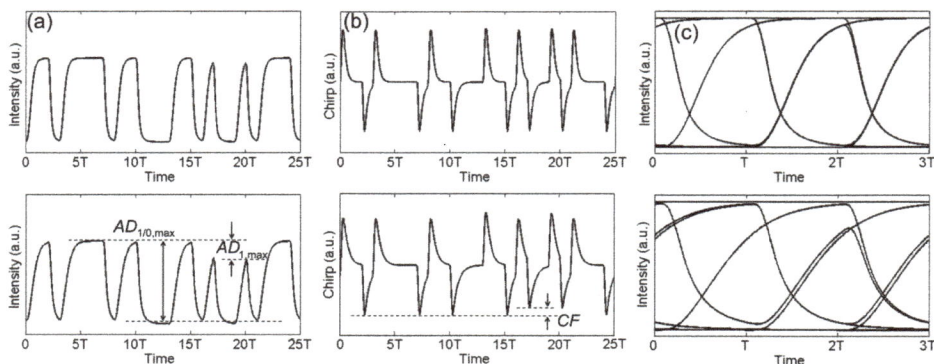

Figure 7. RSOA-encoded pulses (**a**) waveform, (**b**) chirp and (**c**) pseudo-eye diagram at 3 Gb/s (top) and 5 Gb/s (bottom).

A passive single-bus MRR employed as notch filter can allow one to directly modulate the RSOA at enhanced data rate with acceptable performance. To this aim, we must properly select the MRR radius and detuning, and Figure 8 depicts the EP versus these parameters for three different RSOA direct modulation rates which are at least three times higher than the maximum data rate being possible for the RSOA alone. For acceptable performance, not only the EP should lie below the defined FEC limit, but also $AD_{1,max}$ should be below 1 dB [60], $AD_{1/0,max}$ over 10 dB [13], CF should tend to unity, or equivalently 0%, so that the phase variation per time increment of the encoded pulses is as balanced as possible [61], while the overshoot (OS) that inevitably manifests on the encoded pulses [22] should be kept within 25% [62]. The MRR key parameter specifications that efficiently account for, and compromise between, these conditions so that all metrics are acceptable up to 11 Gb/s are $R \in \{8, 12, 16, 20\}$ μm and $\Delta\lambda = 0.67$ nm $\pm$ 0.02 nm. These specifications have been derived by combining information from the results obtained both graphically (EP) and numerically ($AD_{1,max}$, $AD_{1/0,max}$, CF, OS). The existence of a specific permissible range of values for both MRR radius and detuning is physically attributed to the conditions that must be fulfilled for the MRR-based filter to efficiently mitigate the pattern-dependent impairments of the directly modulated RSOA. This requires the MRR to properly act upon the encoded pulses spectrum, which due to the RSOA direct modulation has been modified in accordance with the binary content and position of each pulse. For this purpose, the MRR transmission properties must be suitably tailored, which involves choosing and controlling the wavelength spacing (FSR) as well as the contrast and the position of the notches, by following and fulfilling the necessary conditions. The general guidelines are that: (a) The FSR must be adjusted by taking into account the trade-off between the margin of the TF spectral border, which is defined by the difference between the reference data and the nearest notch wavelength, the TF passband width, the spectral components' suppression degree dependence on pulse peak amplitude they originate from, and the optical carrier level of transmission after filtering; (b) the repetitive notches must be sharp and deep enough to maximize their magnitude difference from their adjacent transmission peaks, which defines the PNCR; and (c) the notches must occur at a longer wavelength than that of the encoded signal so that the MRR transmittance is decreased as the wavelength is increased.

The beneficial effect of an MRR with optimum radius and detuning $R = 20$ μm and $\Delta\lambda = 0.65$ nm, respectively, or $\simeq$ FSR/20, which fall within the range of permissible values specified above, on the encoded signal characteristics at 11 Gb/s is shown in the lower part of Figure 9, where for comparison the same results for the RSOA only are depicted in the upper part. The obtained performance metrics are $EP = 3.56 \times 10^{-4}$ (MRR) vs. 2.86×10^{-1} (RSOA), $AD_{1,max} = 0.64$ dB (MRR) vs. 3.32 dB (RSOA), $AD_{1/0,max} = 10.63$ dB (MRR) vs. 5.88 dB (RSOA), $CF = 9\%$ (MRR) vs. 48% (RSOA), $EO = 87.4\%$ (MRR) vs. 44.5% (RSOA) and $OS = 8\%$. In addition, the net gain (NG) of the RSOA–MRR system [51] is 6.54 dB, thus being exploitable for direct modulation purposes [24]. This value can be increased to approach 10 dB provided that the coupling coefficient, *r*, tends closer to unity, since in this case the creation of the MRR spectral response, according to the general guidelines mentioned above for efficiently mitigating the pattern-dependent impairments of the directly modulated RSOA, is favored. Still, such choice of the specific parameter is practically compromised by the tighter adjustments related to the gap between the straight and bending waveguides. On the other hand, varying the MRR radius from the maximum to the minimum permissible value specified for this parameter does not improve the NG. In fact, this action increases the FSR, which in turn makes the slope of the MRR transfer function less steep. This increases the likelihood that the amplified optical carrier will fail to fall close to the transmission peak, thus suffering by the MRR a greater attenuation of its intensity, which results in less available NG.

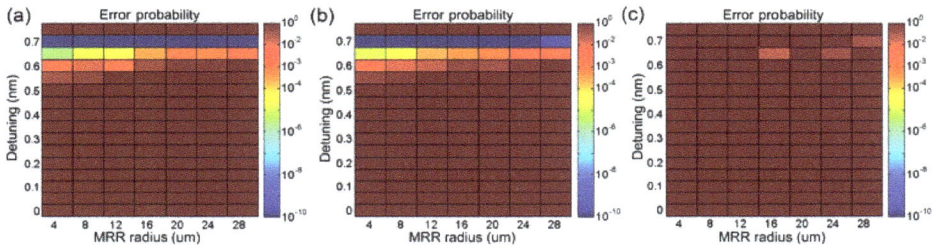

Figure 8. Error probability at MRR output for different MRR radii and detuning at (**a**) 10.5 Gb/s, (**b**) 11 Gb/s and (**c**) 11.5 Gb/s.

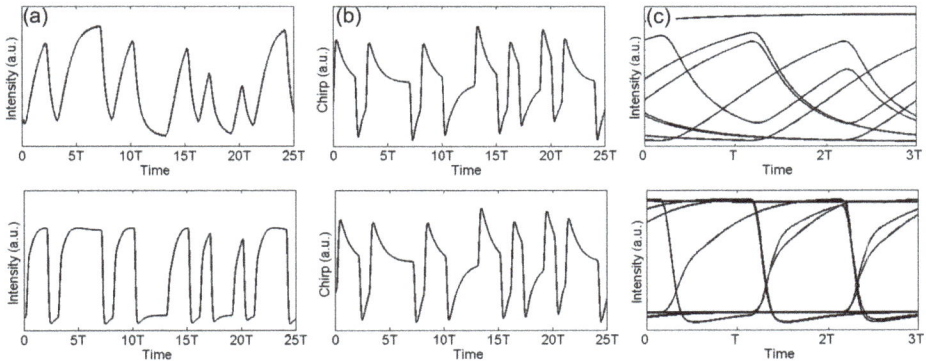

Figure 9. RSOA output (top) and MRR output (bottom) encoded pulses (**a**) waveform, (**b**) chirp and (**c**) pseudo-eye diagram at 11 Gb/s.

Figure 10 shows that owing to the MRR, the RSOA modulation bandwidth is extended indeed to 11 GHz. The peaking observed in the modulation response is induced by the MRR transient time dynamics and is the physical result of interference between the light that circulates inside the ring and the light inserted in the bus waveguide from the RSOA [63]. Consequently, its magnitude should depend on the fraction of the power that comes from the RSOA and enters the ring, that is, $(1 - r^2) \times 100\%$. Our simulations conducted in the critical coupling regime thus show that when this quantity is 9.75%, the frequency position of peaking is shifted to its maximum, which however is not exploitable for RSOA direct modulation at the corresponding data rate due to the constraints imposed by *EP*, $AD_{1,max}$, $AD_{1/0,max}$, *CF* and *OS*. In contrast, when the percentage of coupled power drops to ~2%, the peaking can be leveraged for enhancing the RSOA modulation bandwidth to an extent that is determined approximately by the detuning between the encoded signal and the MRR notch at resonance, that is, $\approx \delta/2\pi$ [24]. On the other hand, according to the above justification for the appearance of peaking, the latter is not affected by changes in the ring radius values, which alter the order of resonance but not the spectral position of the notch. This happens in our case, since the derived permissible values of the MRR radius follow this pattern by being integer multiples of 4 μm. From a mathematical perspective, this leaves intact the parameter δ and accordingly the degree of peaking.

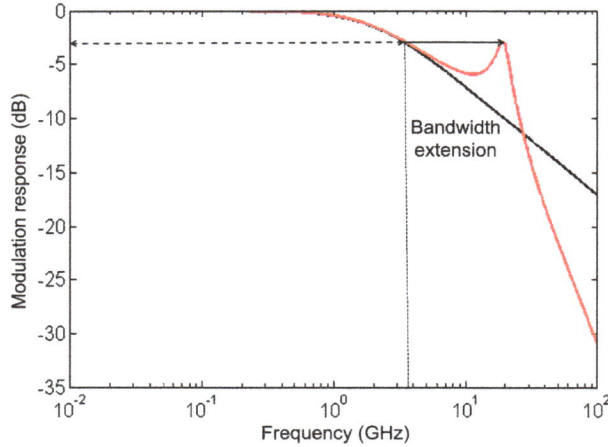

Figure 10. Modulation response of RSOA alone (black line) and with the addition of MRR (red line).

Finally, Figure 11 depicts the MRR transfer function and its impact on the spectral components of the maximum and minimum logical '1's and '0's within the RSOA-encoded data stream. According to [24], it can be seen that the spectral peaks of the encoded logical '1's are shifted to the longer sideband relative to the optical carrier, while those of the encoded logical '0's are shifted to the shorter sideband. In order to combat the deleterious consequences of RSOA direct modulation provoked by the limited RSOA modulation bandwidth, the spectral peaks of the encoded '1's must lie in the falling slope of the MRR TF, with the peak of the maximum '1' being located closer to the TF notch than the peak of the minimum '1', so that the MRR transmits the former less than the latter. Moreover, the spectral peaks of the encoded logical '0's must be confined around the flat portion of the MRR TF, with the peak of the minimum '0' being located nearer to the MRR TF transparency point than the peak of the maximum '0', so that the MRR favors the former more than the latter. These requirements are efficiently satisfied by using the specified optimum parameters of the MRR, whose spectral response figures-of-merit thus are [40] $FSR = \lambda_{enc}^2/(2\pi n_{eff}R) = 13.5$ nm, resonance peak full-width at half-maximum $FWHM = \lambda_{enc}^2(1 - rl)/(2\pi^2 n_{eff}R\sqrt{rl}) = 0.44$ nm and finesse $F = FSR/FWHM = 31$, which have been calculated by substituting $\lambda_{enc} = 1550$ nm, $n_{eff} = 1.41$, $R = 20$ μm, and $r = l = 0.95$ so that the PNCR is ideally infinite. Then, the MRR can act upon the spectral components of the encoded pulses so that the peak differences can become more even for pulses of the same binary content, that is, '1's vs '1's or '0's vs '0's, and more distinguishable between pulses of different binary content, that is, '1's vs '0's. In addition, the absolute magnitudes of marks and spaces can become more enhanced and suppressed, respectively. In this manner, the MRR transforms the pattern-dependent distortions, which have been mapped on the encoded signal's spectral components, into pulse amplitude variations [15] that cancel those present right after the directly modulated RSOA, thus restoring the quality of the encoded pulses.

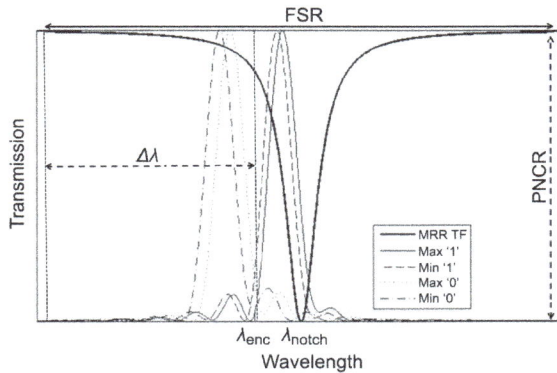

Figure 11. MRR spectral response and action upon spectral components of maximum and minimum encoded logical '1's and '0's at directly modulated RSOA output stream. FSR: free spectral range. PNCR: peak-to-notch contrast ratio. $\Delta\lambda$: detuning (wavelength offset between encoded signal spectral position, λ_{enc}, and shorter sideband transmission peak located FSR/2 away from notch, λ_{notch}).

The MRR target operation wavelengths are those mostly utilized in the applications mentioned at the beginning of the Section 1, that is, in the standard transmission windows of fiber communication systems [57]. Operation at these wavelengths is enabled both by the RSOA broad gain bandwidth and adjustable active medium material composition as well as by the MRR periodic comb-like transmission profile. The latter feature can be exploited provided that the center wavelengths of the encoded data are spaced apart by integer multiples of the free spectral range. In this case, the wavelength allocation should be coarse, in terms of the standardized wavelength division multiplexing grid spacing, or, if a more dense arrangement is desired so as to accommodate more channels, multiple MRRs can be cascaded along the same waveguide bus [64].

The MRR fabrication with regard to (1) size, (2) coupling degree and (3) detuning is technologically feasible according to the following considerations. More specifically: (1) The MRR radius can be down to the very small micrometer scale using high-index contrast waveguide materials [40,65]. (2) The matching between the MRR field transmission coefficient and amplitude attenuation factor for critical coupling is allowed to deviate by 3% from being perfect and still obtain a high PNCR [42]. This is possible by controlling the gap between the straight and bending waveguides using electrically-driven micro-electromechanical system (MEMS) microactuators [66]. (3) The position of the notches relative to the encoded data wavelength can be adjusted by physical means, such as the thermo–optic effect. In this case, electrical heaters are placed in the direct vicinity of the MRR to spectrally shift its resonance [65]. This method is particularly efficient for materials of large positive thermo–optic coefficient, such as silicon, since the required amount of detuning across the FSR can be achieved by supplying an electrical power of the order of a few mWs [65]. Overall, the MRR physical dimensions and tolerances can be determined by modeling the MRR using commercially available simulation software platforms based on coupled mode theory. This procedure is executed for a given waveguide structure material (hence effective index), fabrication technology (hence propagation loss) and incident light (hence wavelength) [67], and has been the subject of research elsewhere, thus being out of the scope of this paper. For example [68], a 20 μm-radius MRR fabricated on a silicon-on-insulator substrate consisting of a 1-μm-thick box layer and 0.45-μm-thick device layer, with waveguide width of 0.5 μm, round-trip propagation losses less than 2 dB/cm and gap between straight and bending waveguide of 250 nm, can efficiently act as notch filter, with PNCR 15 dB, on optical signals centered in the vicinity of 1550 nm.

4. Conclusions

In conclusion, we have demonstrated the capability of an MRR configured as notch filter to bypass the limited modulation bandwidth of an RSOA and directly modulate the RSOA nearly three-and-a-half times faster than possible without the MRR. The analysis of the results obtained through modeling and simulation of the RSOA and MRR response reveals that if the MRR critical parameters are properly selected, which is technologically feasible, then the MRR allows one to enhance the RSOA modulation bandwidth with improved performance and encoded signal characteristics. Employing the MRR in practice allows one to leverage the benefits of free spectral range fine adjustment, controllable finesse, sharp spectral selectivity, feasible and versatile detuning, bandwidth and wavelength tunability, enhanced peak-to-notch contrast ratio, the existence of many different material systems and corresponding fabrication processes, integration compatibility with other photonic devices and platforms, and off-the-shelf availability.

Author Contributions: Zoe V. Rizou and Kyriakos E. Zoiros conceived the paper theme, applied and adapted RSOA and MRR modeling, ran the simulations, derived and interpreted the obtained results and wrote the manuscript.

Conflicts of Interest: The authors declare no conflict of interest.

References

1. Spiekman, L.H. Active devices in passive optical networks. *J. Lightwave Technol.* **2013**, *31*, 488–497.
2. Liu, Z.; Sadeghi, M.; de Valicourt, G.; Brenot, R.; Violas, M. Experimental validation of a reflective semiconductor optical amplifier model used as a modulator in radio over fiber systems. *IEEE Photonics Technol. Lett.* **2011**, *23*, 576–578.
3. Meehan, A.; Connelly, M.J. Experimental characterization and modeling of the improved low frequency response of a current modulated bulk RSOA slow light based microwave phase shifter. *Opt. Commun.* **2015**, *341*, 241–244.
4. Guiying, J.; Lirong, H. Remodulation scheme based on a two-section reflective SOA. *J. Semicond.* **2014**, *35*, 054008.
5. Huang, L.; Hong, W.; Jiang, G. All-optical power equalization based on a two-section reflective semiconductor optical amplifier. *Opt. Express* **2013**, *21*, 4598–4611.
6. Peng, P.C.; Shiu, K.C.; Liu, W.C.; Chen, K.J.; Lu, H.H. A fiber-optical cable television system using a reflective semiconductor optical amplifier. *Laser Phys.* **2013**, *23*, 025106.
7. Connelly, M.J. *Semiconductor Optical Amplifiers*; Kluwer Academic Publishers: Dordrecht, The Netherlands, 2002.
8. König, S. *Semiconductor Optical Amplifiers and mm-Wave Wireless Links for Converged Access Networks*; Karlsruhe Series in Photonics & Communications; Karlsruhe Institute of Technology, Institute of Photonics and Quantum Electronics (IPQ): Karlsruhe, Germany, 2014; Volume 14.
9. Cho, K.Y.; Takushima, Y.; Chung, Y.C. 10-Gb/s operation of RSOA for WDM PON. *IEEE Photonics Technol. Lett.* **2008**, *20*, 1533–1535.
10. Wei, J.L.; Hamié, A.; Gidding, R.P.; Hugues-Salas, E.; Zheng, X.; Mansoor, S.; Tang, J.M. Adaptively modulated optical OFDM modems utilizing RSOAs as intensity modulators in IMDD SMF transmission systems. *Opt. Express* **2010**, *18*, 8556–8573.
11. Shim, H.K.; Kim, H.; Chung, Y.C. 20-Gb/s polar RZ 4-PAM transmission over 20-km SSMF using RSOA and direct detection. *IEEE Photonics Technol. Lett.* **2015**, *27*, 1116–1119.
12. Cho, K.Y.; Choi, B.S.; Takushima, Y.; Chung, Y.C. 25.78-Gb/s operation of RSOA for next-generation optical access networks. *IEEE Photonics Technol. Lett.* **2011**, *23*, 495–497.
13. Schrenk, B.; de Valicourt, G.; Omella, M.; Lazaro, J.A.; Brenot, R.; Prat, J. Direct 10-Gb/s modulation of a single-section RSOA in PONs with high optical budget. *IEEE Photonics Technol. Lett.* **2010**, *22*, 392–394.
14. Rizou, Z.V.; Zoiros, K.E.; Morel, P. Improving SOA direct modulation capability with optical filtering. In Proceedings of the International Conference on Transparent Optical Networks, Trento, Italy, 10–14 July 2016.
15. Papagiannakis, I.; Omella, M.; Klonidis, D.; Birbas, A.N.; Kikidis, J.; Tomkos, I.; Prat, J. Investigation of 10-Gb/s RSOA-based upstream transmission in WDM-PONs utilizing optical filtering and electronic equalization. *IEEE Photonics Technol. Lett.* **2008**, *20*, 2168–2170.

16. Cossu, G.; Bottoni, F.; Corsini, R.; Presi, M.; Ciaramella, E. 40 Gb/s single R-SOA transmission by optical equalization and adaptive OFDM. *IEEE Photonics Technol. Lett.* **2013**, *25*, 2119–2122.

17. Kim, H. 10-Gb/s operation of RSOA using a delay interferometer. *IEEE Photonics Technol. Lett.* **2010**, *22*, 1379–1381.

18. Su, T.; Zhang, M.; Chen, X.; Zhang, Z.; Liu, M.; Liu, L.; Huang, S. Improved 10-Gbps uplink transmission in WDM-PON with RSOA-based colorless ONUs and MZI-based equalizers. *Opt. Laser Technol.* **2013**, *51*, 90–97.

19. Zhang, M.; Wang, D.; Cao, Z.; Chen, X.; Huang, S. Suppression of pattern dependence in 10 Gbps upstream transmission of WDM-PON with RSOA-based ONUs. *Opt. Commun.* **2013**, *308*, 248–252.

20. Presi, M.; Chiuchiarelli, A.; Corsini, R.; Choudury, P.; Bottoni, F.; Giorgi, L.; Ciaramella, E. Enhanced 10 Gb/s operations of directly modulated reflective semiconductor optical amplifiers without electronic equalization. *Opt. Express* **2012**, *20*, B507–B512.

21. Zoiros, K.E.; Morel, P. Enhanced performance of semiconductor optical amplifier at high direct modulation speed with birefringent fiber loop. *AIP Adv.* **2014**, *4*, 077107.

22. Engel, T.; Rizou, Z.V.; Zoiros, K.E.; Morel, P. Semiconductor optical amplifier direct modulation with double-stage birefringent fiber loop. *Appl. Phys. B-Lasers Opt.* **2016**, *122*, 158.

23. Zoiros, K.E.; Morel, P.; Hamze, M. Performance improvement of directly modulated semiconductor optical amplifier with filter-assisted birefringent fiber loop. *Microw. Opt. Technol. Lett.* **2015**, *57*, 2247–2251.

24. Rizou, Z.V.; Zoiros, K.E. Performance analysis and improvement of semiconductor optical amplifier direct modulation with assistance of microring resonator notch filter. *Opt. Quantum Electron.* **2017**, *49*, 119.

25. Katz, O.; Malka, D. Design of novel SOI 1 × 4 optical power splitter using seven horizontally slotted waveguides. *Photonics Nanostruct.* **2017**, *25*, 9–13.

26. Malka, D.; Cohen, M.; Turkiewicz, J.; Zalevsky, Z. Optical micro-multi-racetrack resonator filter based on SOI waveguides. *Photonics Nanostruct.* **2015**, *16*, 16–23.

27. Antonelli, C.; Mecozzi, A.; Hu, Z.; Santagiustina, M. Analytic study of the modulation response of reflective semiconductor optical amplifiers. *J. Lightwave Technol.* **2015**, *33*, 4367–4376.

28. Stathi, G.; Rizou, Z.V.; Zoiros, K.E. Simulation of directly modulated RSOA. In Proceedings of the International Conference on Numerical Simulation of Optoelectronic Devices, Copenhagen, Denmark, 24–28 July 2017.

29. Connelly, M.J. Reflective semiconductor optical amplifier pulse propagation model. *IEEE Photonics Technol. Lett.* **2012**, *24*, 95–97.

30. Sengupta, I.; Barman, A.D. Analysis of optical re-modulation by multistage modeling of RSOA. *Optik* **2014**, *125*, 3393–3400.

31. Cassioli, D.; Scotti, S.; Mecozzi, A. A time-domain computer simulator of the nonlinear response of semiconductor optical amplifiers. *IEEE J. Quantum Electron.* **2000**, *36*, 1072–1080.

32. Agrawal, G.P.; Olsson, N.A. Self-phase modulation and spectral broadening of optical pulses in semiconductor laser amplifiers. *IEEE J. Quantum Electron.* **1989**, *25*, 2297–2306.

33. Zhou, E.; Zhang, X.; Huang, D. Analysis on dynamic characteristics of semiconductor optical amplifiers with certain facet reflection based on detailed wideband model. *Opt. Express* **2007**, *15*, 9096–9106.

34. Antonelli, C.; Mecozzi, A. Reduced model for the nonlinear response of reflective semiconductor optical amplifiers. *IEEE Photonics Technol. Lett.* **2013**, *25*, 2243–2246.

35. Shen, T.M.; Agrawal, G.P. Pulse-shape effects on frequency chirping in single-frequency semiconductor lasers under current modulation. *J. Lightwave Technol.* **1986**, *4*, 497–503.

36. Cartledge, J.C.; Burley, G.S. The effect of laser chirping on lightwave system performance. *J. Lightwave Technol.* **1989**, *7*, 568–573.

37. Zoiros, K.E.; Botsiaris, C.; Koukourlis, C.S.; Houbavlis, T. Necessary temporal condition for optimizing the switching window of the semiconductor-optical-amplifier-based ultrafast nonlinear interferometer in counter-propagating configuration. *Opt. Eng.* **2006**, *45*, 115005.

38. Ali, M.A.; Elrefaie, A.F.; Ahmed, S.A. Simulation of 12.5 Gb/s lightwave optical time-division multiplexer using semiconductor optical amplifiers as external modulators. *IEEE Photonics Technol. Lett.* **1992**, *4*, 280–283.

39. Chi, J.W.D.; Chao, L.; Rao, M.K. Time-domain large-signal investigation on nonlinear interactions between an optical pulse and semiconductor waveguides. *IEEE J. Quantum Electron.* **2001**, *37*, 1329–1336.

40. Rabus, D.G. *Integrated Ring Resonators: The Compendium*; Spring: Berlin, Germany, 2007.

41. Heebner, J.E.; Wong, V.; Schweinsberg, A.; Boyd, R.W.; Jackson, D.J. Optical transmission characteristics of fiber ring resonators. *IEEE J. Quantum Electron.* **2004**, *40*, 726–730.
42. Absil, P.P.; Hryniewicz, J.V.; Little, B.E.; Wilson, R.A.; Joneckis, L.G.; Ho, P.T. Compact microring notch filters. *IEEE Photonics Technol. Lett.* **2000**, *12*, 398–400.
43. Nielsen, M.L. Experimental and Theoretical Investigation of Semiconductor Optical Amplifier (SOA) Based All-Optical Switches. Ph.D. Thesis, Technical University of Denmark, Lyngby, Denmark, 2004.
44. Pile, B.; Taylor, G. Small-signal analysis of microring resonator modulators. *Opt. Express* **2014**, *22*, 14913–14928.
45. Sato, K.; Toba, H. Reduction of mode partition noise by using semiconductor optical amplifiers. *IEEE J. Sel. Top. Quantum Electron.* **2001**, *7*, 328–333.
46. Ahmed, M.; Lafi, A.E. Analysis of small-signal intensity modulation of semiconductor lasers taking account of gain suppression. *Pramana-J. Phys.* **2008**, *71*, 99–115.
47. Corless, R.M.; Gonnet, G.H.; Hare, D.E.; Jeffrey, D.J.; Knuth, D.E. On the Lambert W function. *Adv. Comput. Math.* **1996**, *5*, 329–359.
48. Dawson, H.G. On the Numerical Value of $\int_0^h e^{x^2} dx$. *Proc. Lond. Math. Soc.* **1897**, *1*, 519–522.
49. Totović, A.R.; Crnjanski, J.V.; Krstić, M.M.; Gvozdić, D.M. Numerical study of the small-signal modulation bandwidth of reflective and traveling-wave SOAs. *J. Lightwave Technol.* **2015**, *33*, 2758–2764.
50. De Valicourt, G.; Pommereau, F.; Poingt, F.; Lamponi, M.; Duan, G.; Chanclou, P.; Violas, M.; Brenot, R. Chirp reduction in directly modulated multi-electrode RSOA devices in passive optical networks. *IEEE Photonics Technol. Lett.* **2010**, *22*, 1425–1427.
51. Zoiros, K.E.; Rizou, Z.V.; Connelly, M.J. On the compensation of chirp induced from semiconductor optical amplifier on RZ data using optical delay interferometer. *Opt. Commun.* **2011**, *284*, 3539–3547.
52. Liu, F.; Wang, T.; Qiang, L.; Ye, T.; Zhang, Z.; Qiu, M.; Su, Y. Compact optical temporal differentiator based on silicon microring resonator. *Opt. Express* **2008**, *16*, 15880–15886.
53. Trebino, R. *Frequency-Resolved Optical Gating: The Measurement of Ultrashort Laser Pulses*; Springer Science & Business Media: New York, NY, USA, 2012; pp. 11–35.
54. Azaña, J.; Park, Y.; Li, F. Linear self-referenced complex-field characterization of fast optical signals using photonic differentiation. *Opt. Commun.* **2011**, *284*, 3772–3784.
55. Watts, R.T.; Shi, K.; Barry, L.P. Time-resolved chirp measurement for 100GBaud test systems using an ideal frequency discriminator. *Opt. Commun.* **2012**, *285*, 2039–2043.
56. Ruege, A.C. Electro-Optic Ring Resonators in Integrated Optics for Miniature Electric Field Sensors. Ph.D. Thesis, The Ohio State University, Columbus, OH, USA, 2001.
57. Agrawal, G.P. *Fiber-Optic Communication Systems*; Wiley: New York, NY, USA, 2002.
58. Wei, J.L.; Hamié, A.; Giddings, R.P.; Tang, J.M. Semiconductor optical amplifier-enabled intensity modulation of adaptively modulated optical OFDM signals in SMF-based IMDD systems. *J. Lightwave Technol.* **2009**, *27*, 3678–3688.
59. Rizou, Z.V.; Zoiros, K.E.; Hatziefremidis, A. Signal amplitude and phase equalization technique for free space optical communications. In Proceedings of the International Conference on Transparent Optical Networks, Cartagena, Spain, 23–27 July 2013.
60. Vardakas, J.S.; Zoiros, K.E. Performance investigation of all-optical clock recovery circuit based on Fabry-Pérot filter and semiconductor optical amplifier assisted Sagnac switch. *Opt. Eng.* **2007**, *46*, 085005.
61. Zoiros, K.E.; Vardakas, J.S.; Tsigkas, M. Study on the instantaneous frequency deviation of pulses switched from semiconductor optical amplifier–assisted Sagnac interferometer. *Opt. Eng.* **2010**, *49*, 075003.
62. Hinton, K.; Stephens, T. Modeling high-speed optical transmission systems. *IEEE J. Sel. Area Commun.* **1993**, *11*, 380–392.
63. Müller, J.; Merget, F.; Azadeh, S.S.; Hauck, J.; García, S.R.; Shen, B.; Witzens, J. Optical peaking enhancement in high-speed ring modulators. *Sci. Rep.* **2014**, *4*, 6310.
64. Padmaraju, K.; Bergman, K. Resolving the thermal challenges for silicon microring resonator devices. *Nanophotonics* **2014**, *3*, 269–281.
65. Bogaerts, W.; De Heyn, P.; Van Vaerenbergh, T.; De Vos, K.; Kumar Selvaraja, S.; Claes, T.; Dumon, P.; Bienstman, P.; Van Thourhout, D.; Baets, R. Silicon microring resonators. *Laser Photonics Rev.* **2012**, *6*, 47–73.
66. Lee, M.C.; Wu, M.C. MEMS-actuated microdisk resonators with variable power coupling ratios. *IEEE Photonics Technol. Lett.* **2005**, *17*, 1034–1036.

67. Stoffer, R.; Hiremath, K.R.; Hammer, M.; Prkna, L.; Čtyroký, J. Cylindrical integrated optical microresonators: Modeling by 3-D vectorial coupled mode theory. *Opt. Commun.* **2005**, *256*, 46–67.
68. Niehusmann, J.; Vörckel, A.; Bolivar, P.H.; Wahlbrink, T.; Henschel, W.; Kurz, H. Ultrahigh-quality-factor silicon-on-insulator microring resonator. *Opt. Lett.* **2004**, *29*, 2861–2863.

applied
sciences

MDPI

Article

SOA Based Photonic Integrated WDM Cross-Connects for Optical Metro-Access Networks

Nicola Calabretta *, Wang Miao, Ketemaw Mekonnen and Kristif Prifti

IPI—Institute of Photonic Integration, Eindhoven University of Technology, 5600 MB Eindhoven,
The Netherlands; w.miao@tue.nl (W.M.); K.A.Mekonnen@tue.nl (K.M.); kristifprifti@hotmail.com (K.P.)
* Correspondence: n.calabretta@tue.nl; Tel.: +31-40-247-5361

Received: 30 July 2017; Accepted: 18 August 2017; Published: 23 August 2017

Abstract: We present a novel optical metro node architecture that exploits the Wavelength Division Multiplexing (WDM) optical cross-connect nodes for interconnecting network elements, as well as computing and storage resources. The photonic WDM cross-connect node based on semiconductor optical amplifiers (SOA) allows switching data signals in wavelength, space, and time for fully exploiting statistical multiplexing. The advantages of using an SOA to realize the WDM cross-connect switch in terms of transparency, switching speed, photonic integrated amplification for loss-less operation, and gain equalization are verified experimentally. The experimental assessment of a 4 × 4 photonic integrated WDM cross-connect confirmed the capability of the cross-connect chip to switch the WDM signal in space and wavelength. Experimental results show lossless operation, low cross-talk <−30 dB, and dynamically switch within few nanoseconds. Moreover, the operation of the cross-connect switch with multiple WDM channels and diverse modulation formats is also investigated and reported. Error-free operation with less than a 2 dB power penalty for a single channel, as well as WDM input operation, has been measured for multiple 10/20/40 Gb/s NRZ-OOK, 20 Gb/s PAM4, and data-rate adaptive DMT traffic. Compensation of the losses indicates that the modular architecture could scale to a larger number of ports.

Keywords: semiconductor optical amplifiers; photonic integrated cross-connect switch; wavelength division multiplexing; optical metropolitan networks

1. Introduction

Optical metro networks face significant challenges supporting ever-increasing bandwidth demands and ever increasing service expectations [1]. High-performance next-generation dynamic optical metro networks should efficiently support a variety of access applications with dynamic traffic patterns (LTE and 5G backhaul and fronthaul, multi-technology Passive Optical Networks (PON), data center interconnects, enterprises, etc.) as well as multi-Tbit/s interfaces with core networks by leveraging the latest advances in optical transmission and switching. Moreover, applications, such as 5G with deployment of multiple antennas and MIMO radio configurations, require not only large bandwidth beyond 100 Gb/s, but also the computing and storage resources for processing the signals from the radio antennas. Therefore, next generation metropolitan nodes will co-allocate telecom network elements and compute and storage resources to cope with such applications. In addition, new developments in network virtualization could partition the optical data layer to be able to accommodate a wide range of use cases, from the vertical industries and other infrastructure users with different requirements (e.g., latency, resiliency, and bandwidth) allocating logical networks and infrastructures, optimally tailored for each specific use case. This requires the metro node network architecture to be a flexible infrastructure that can be adapted and scaled on demand according to the applications. On the other hand, power consumption and costs of such infrastructures should be sustainable as the infrastructure scales with data rate and network elements, as well as computing and storage resources.

Several programmable optical metro node architectures based on simplified, flexible, and cost-efficient optical switching and transmission technology, with high capacity and faster network reconfiguration and with efficient exploitation of the available optical spectrum, have been currently investigated [2–6]. Novel node architectures based on the "whitebox" concept, including "distributed DC" capabilities, i.e., using disaggregated hardware in a vendor-neutral approach and having local processing and storage resources, is highly promising for integrating commodity hardware with telecom network elements. In the target architecture of Central Office Re-architected as a Datacenter (CORD) [3] shown in Figure 1, it includes a collection of commodity servers interconnected by a fabric constructed from electrical whitebox switches. The switching fabric is organized in a leaf and spine topology to optimize for traffic flowing east to west between the access network that connects customers to the Central Office and the upstream links that connect the Central Office to the operator's core network.

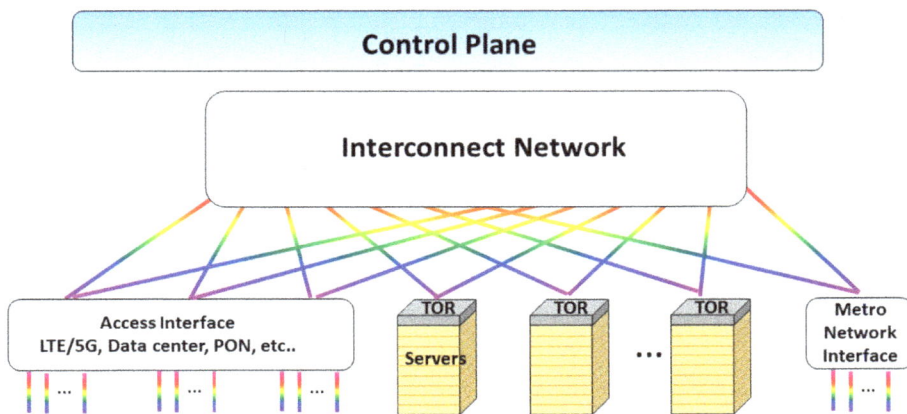

Figure 1. Target hardware built from commodity servers, I/O blades, and switches.

Relying on the implementation of a high-bandwidth electronic switch node is limited by the switch ASIC I/O bandwidth due to the scaling issues of the ball grid array (BGA) package [7]. Higher bandwidth is achievable by stacking several ASICs in a multi-tier structure, but at the expense of larger latency, higher cost, and power consumption. Moreover, power consumption is an additional issue for such switches. In addition to the power-hungry electronic switching fabrics, the E/O and O/E conversions at the switching node actually dissipate a large portion of the consumed power [8].

Switching the data signals in the optical domain has the potential to overcome the scaling issues of electronic switches [9]. The advantages of deploying optical switching technologies in an optical metro node are multi-fold:

Transparency to data-rate and data-format allows for extremely high I/O bandwidth without implementing signal-dependent interfaces.

(1) The high capacity helps to flatten the network topology, avoiding bandwidth bottleneck and large latency caused by hierarchical structures.

(2) Massive O/E/O conversions can be eliminated, improving the energy-efficiency and cost-efficiency.

(3) The reduced number of cables benefitting from the high port capacity may facilitate the deployment and management of the network.

Several optical switching schemes have been proposed and investigated [10,11]. Optical switches based only on space have inefficient bandwidth utilization and inflexible connection. On the other hand, only time domain fast optical switches can offer sub-wavelength granularity for on-demand and high-degree connectivity by exploiting statistical multiplexing. To fully reap the profits from the statistical multiplexing, the switch should support fast scheduling and reconfiguration at the nanosecond scale. Nevertheless, the optical switch featuring with nanoseconds reconfiguration time has been only demonstrated with limited port-count [12–14]. An innovative optical interconnect network architecture for data center network architecture based on fast flow-controlled optical cross-connect (OXC) switches have been employed as the distributed switching elements in [15]. The capability of switching the aggregated traffic in both time and wavelength domain at the nanosecond scale has further improved the flexibility and feasibility of the system. However, no application to optical metro network node architecture has been reported, and experimental assessment of the WDM optical cross-connect switch has been performed only for non-return to zero signals and for a limited optical power dynamic range.

In this work we present a novel optical metro node architecture that exploits the WDM optical cross-connect nodes for interconnecting network elements, as well as computing and storage resources. The experimental assessment of a photonic WDM cross-connect node based on semiconductor optical amplifiers (SOA) as a main building functionality for implementing the interconnected network of the metro node is also reported. The advantages of using an SOA to realize the WDM cross-connect switch in terms of switching speed, photonic integrated amplification for loss-less operation, and gain equalization will be verified experimentally. Experimental results on the assessment of a 4 × 4 photonic integrated WDM cross-connect confirmed the capability of the cross-connect chip to switch WDM signal in space and wavelength. Experimental results show lossless operation and cross-talk < −30 dB. Compensation of the losses is a good indication that the modular architecture could scale to a larger number of ports. Moreover, the operation of the cross-connect switch with multiple WDM channels and diverse modulation formats is also investigated. Experimental results confirmed the capability of the cross-connect chip to dynamically switch, within a few nanoseconds, the WDM data packets in space and wavelength. Error-free operation with less than a 2 dB power penalty for a single channel, as well as WDM input operation, has been measured for multiple 40 Gb/s NRZ-OOK, 20 Gb/s PAM4, and data-rate adaptive DMT traffic.

The paper is organized as follow. In Section 2, a brief description of the optical metro node architecture comprising of the WDM optical cross-connects based on SOAs will be presented. The design and operation principle of the SOA-based photonic WDM cross-connect will be discussed in detail in Section 3. The photonic WDM cross-connect chip fabrication and characterization will be described in Section 4. The experimental assessment of the photonic switch under single and multiple channel operation, and multiple modulation formats, will be reported in Section 5. Finally, conclusions will be summarized in Section 6.

2. Optical Metro Node Architecture Based on the WDM Cross-Connect

The optical metro node architecture exploiting the WDM cross-connects is reported in Figure 2. The optical node architecture include both optical and electrical (interfaces) switching. It consists of multiple interfaces (PON, LTE/5G, enterprise, metro network) that use multi-protocol, multilink interfaces, and hybrid electronic/optical switches to interface with access network segments along with the ability to mix and match protocols (i.e., CPRI, IP, Ethernet, etc.) over the same physical link. Moreover, the optical node includes also a number of servers organized in racks and interconnected via an electronic Top-of-the-Rack (TOR) switch. Each interface and TOR switch is equipped with a number of WDM bi-directional optical links. As shown in Figure 2, the interfaces and TORs are interconnected by a number of WDM cross-connect in a spine and leaf architecture. The main difference with the CORD approach is that the interconnect networks is built on a WDM cross-connect instead of electronic switches as in CORD. This allows to transparently switch the data traffic between the interfaces and the TORs without expensive O/E/O converters, greatly reducing the power consumption and costs. Moreover, the ability of the WDM cross-connect to switch data signals not only in wavelength and space, but also in time (at the nanosecond time scale), allows to fully exploit the statistical multiplexing switching at the optical data plane level. The number of interconnected interfaces and TORs and the capacity depends on the radix of the WDM cross-connect and the amount of parallel WDM cross-connect. For the PON, LTE/5G, and enterprise, the interface consists of the access interface and network interface, while for TOR it consists of a server interface and the network interface. Part of the traffic that belongs to the same access (the same PON or same server pool in the same rack) is then directly exchanged by the interfaces (or TORs). The other part of the traffic is directed to the network interface via multiple WDM transceivers, eventually with variable data rate and modulation formats. The reason for employing multiple WDM transceivers is that WDM allows scaling the communication bandwidth between the interfaces/TORs and the WDM cross-connect network employing multiple wavelengths to generate a high capacity channel. The control plane is, therefore, in charge of (re-)configuring the transceivers (data format and data rate) and, accordingly, the WDM cross-connects. As the main topic of this work is not the control plane and the implementation of the interfaces, the rest of the paper will discuss in detail and assess the design, photonic integration, and validation of the SOA-based WDM cross-connects.

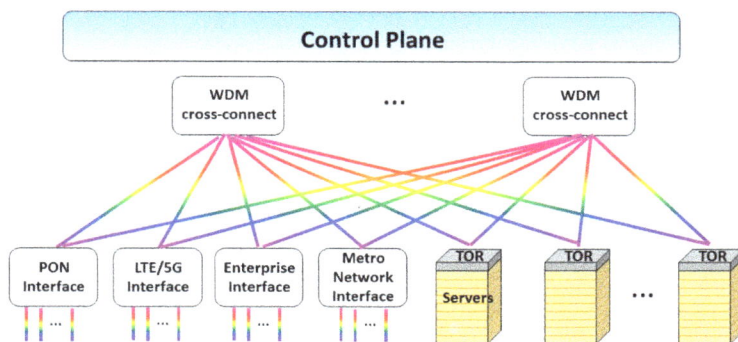

Figure 2. Optical metro node architecture exploiting the Wavelength Division Multiplexing (WDM) cross-connects.

3. SOA-Based WDM Cross-Connects

The schematic of the optical wavelength, space, and time cross-connect switch is illustrated in Figure 3. The non-blocking optical cross-connect has N inputs, and each input carries M different wavelengths generated by the interfaces/TORs. The modular cross-connect processes the N WDM inputs in parallel by the respective optical modules, and forwards the individual wavelength channels to any output ports according to the switching control signals provided by the control plane. A possible solution for fast (sub-microseconds) controlling of the switch is reported in [16]. Each optical module consists of a 1:N splitter to broadcast the WDM channels to the N wavelength selective switches (WSS). The outputs of the N WSSs are connected to the N wavelength combiners of the respective N output ports. Each WSS can select one or more wavelength channels and forward the channels to the output ports according to the control signals. The WSS consists of two AWGs (acting as a demultiplexer and a multiplexer) and M SOA based optical gates. The first 1xM AWG operates as wavelength demultiplexer. Turning on or off the M SOA optical gates determines which wavelength channel is forwarded to the output or is blocked. The second Mx1 AWG operates as a wavelength multiplexer. Multicast operation is also possible with this architecture. The broadband operation of the SOA enables the selection of any wavelength in the C band. Moreover, the amplification provided by the SOA compensates the losses introduced by the two AWGs. It should be noted that the amount of SOAs is NxNxM, but typically only NxM will be turned on. Moreover, during the switching operation, the modules operate in an independent and parallel way to each other, which introduces important features such as distributed control for the optical switch, leading to the following advantages. Firstly, the overall performance of the optical switch can be evaluated by testing a single module. Secondly, the parallel and independent operation of the module make the control complexity and the switching time (latency) of the entire switch independent of the port count and equal to the switching time of a single module. Furthermore, scaling the port count leads to a linear increase in components and energy consumption, by employing copies of the identical modules.

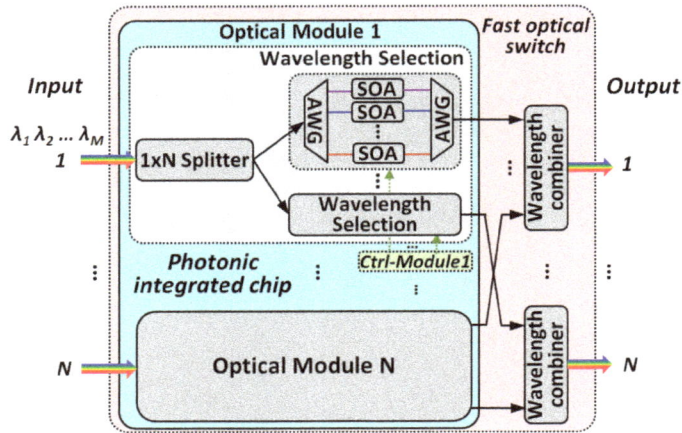

Figure 3. Schematic of the WDM cross-connect based on Semiconductor Optical Amplifiers (SOAs).

4. Photonic Integrated WDM Cross-Connect

Based on the schematic shown in Figure 3, a PIC integrating four optical modules each with four WDM channels (without a wavelength combiner) are designed, as shown in Figure 4. The chip has been realized in a multi-project wafer (MPW) in the Jeppix platform with limited space of the cell (6 mm × 4 mm). Each of the four identical modules processes one of the four WDM inputs. At the input of each module, an 800 μm booster SOA is employed to compensate the 6 dB losses of the 1:4 splitter and, partially, the AWG losses at the WSS. The passive 1:4 splitter is realized by a cascading 1 × 2 multimode interferometer (MMI), with the outputs connected to four identical WSSs, respectively. The AWGs of the WSS are designed with a free spectral range (FSR) of 15 nm, which has been tailored to fit the limited cell size offered in the MWP. The quantum well active InGaAsP/InP SOA gates have a length of 350 μm. The input and output facets of the chip are anti-reflection coated. The chip includes a total of 112 elements (four booster SOA pre-amplifiers with DC bias, 64 gate SOAs with RF bias, 32 AWGs, and 12 1 × 2 MMI couplers). The light-shaded electrodes are wire bonded to the neighboring PCBs to enable the control of the SOA gates. Lensed fibers have been employed to couple the light in and out of the chip. Spectral tuning of the AWGs is, in principle, possible by using heaters on top of the AWG waveguides with a limited tuning range within the channel (not implemented in this design).

The parameters and static performances, including the operational bandwidth of the SOAs, central wavelength, as well as the passband for the AWGs, and crosstalk between different channels have been characterized by employing the experimental setup shown in Figure 5. Four optical input channels spaced by 500 GHz, from $\lambda 1 = 1532$ nm to $\lambda 4 = 1544$ nm, are launched into input port 1 of the photonic chip to characterize the chip. As the optical modules are identical, we have assessed the operation of one optical module to characterize the switching operation. The optical power of the WDM input channels was 2 dBm/channel (see also Figure 5). The input SOA was biased with 100 mA of current. The temperature of the chip was maintained at 25 °C. A polarization controller was employed at the input of the chip due to the high polarization-dependent loss of the quantum well SOAs.

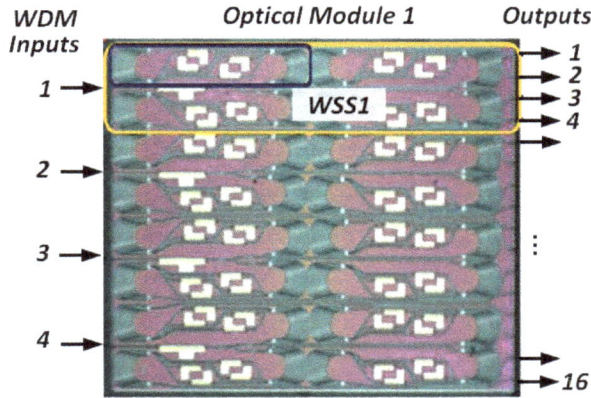

Figure 4. Schematic of the fabricated 4 × 4 fast optical switch PIC.

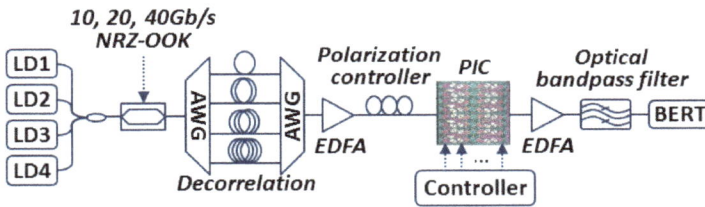

Figure 5. Experimental setup employed to assess the performance of the WDM photonic cross-connect.

First we assess the operation of the single WSS. One WDM channel at a time is statically switched at the WSS output by enabling one of the four SOA gates at a time. The current applied at each of the SOA gates was 62 mA. The four measured spectra at the WSS output (output 1 of the chip) are shown in Figure 6. An on/off switch ratio higher than 30 dB was measured. The optical power at the chip output for Channel 1 and Channel 3 was −9 dBm. Considering 6 dB/facet coupling loss, a 1 dB on chip gain is estimated. However, Channels 2 and 4 have 10 dB of extra losses. Inspection of the chip reveals that two waveguides after the AWG demux are not fully resolved, which leads to substantial optical power coupling between the channels. The coupled signals at different wavelengths are filtered by the output AWG, explaining the extra losses. This can be solved in the next fabrication run.

In the second experiment, Channel 1 has been switched to one of the output ports at a time by enabling one of the SOA gates of the four WSSs at a time. To measure the cross-talk between the WSSs output ports (here we define the cross-talk as the ratio between the optical power of a single channel at the desired output port and the optical power at the other output ports), Figure 7 shows the optical spectra recorded at the four outputs when Channel 1 is only switched at output 1. A cross-talk <−30 dB was measured.

Furthermore, an experiment to validate the capability of the switch to forward multiple wavelengths in parallel was performed. In the setup, two SOA gates of the WSS2 for Channel 1 and Channel 3 are enabled to allow for both channels to be forwarded at output port 2. Figure 8 shows the recorded spectra at the output 2, clearly illustrating the forwarding of Channel 1 and Channel 3, while Channel 2 and Channel 4 are blocked, with a contrast ratio above 30 dB.

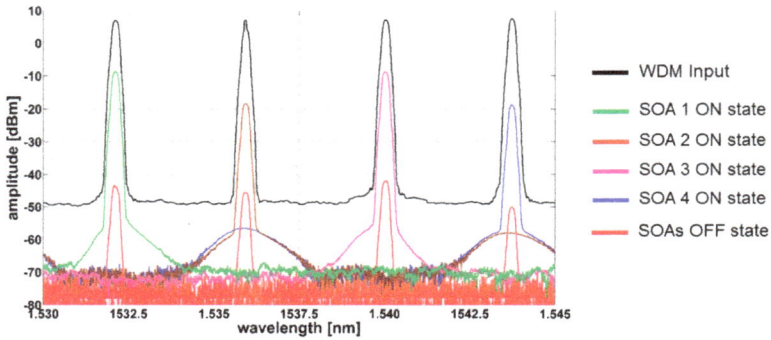

Figure 6. Optical spectra of the WDM channels switched by the Wavelength Selective Switch (WSS).

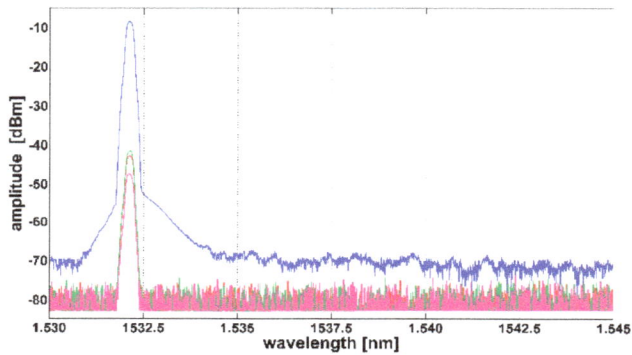

Figure 7. Recorded spectra of Channel 1 at the four WSS outputs (only the SOA gate of WSS1 was enabled).

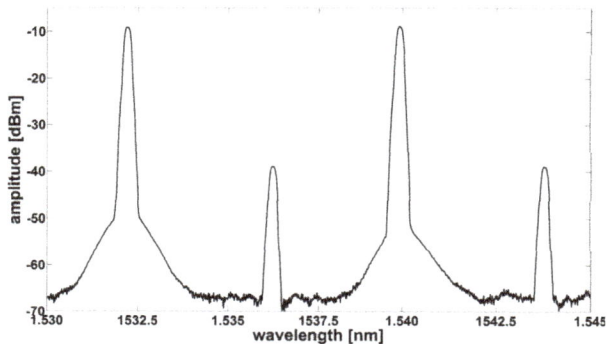

Figure 8. Optical spectra recorded at output port 2 showing the WDM switching operation of the photonic chip.

5. Experimental Results

To assess the performance of the 4 × 4 fast optical switch PIC, the experiment setup shown in Figure 5 is employed. Four WDM optical channels with packetized NRZ-OOK (PRBS 211-1) payloads at data rates of 10 Gb/s, 20 Gb/s, and 40 Gb/s are generated at 1525.0 nm, 1528.9 nm, 1532.9 nm, and

1536.8 nm. The packet has the duration of 540 ns and 60 ns guard time. The four WDM channels are de-correlated, amplified, and injected into the photonic-integrated WDM optical switch PIC via a polarization controller. Module 1, as one of the four identical modules, has been selected for the switching performance assessment. The total power launched into the input Port 1 is 2 dBm (corresponding to −4 dBm per channel). The input SOA, acting as a booster amplifier, is continuously biased with 100 mA of current. The shorter SOAs in the WSS acting as optical gates on the different channels are controlled by an FPGA-based switch controller.

5.1. Dynamic Switching

The dynamic switching operation of the single WSS (WSS1 in Figure 4) is first investigated. The WDM packets arriving at WSS 1 in Module 1 are de-multiplexed and controlled individually by the gate SOAs. The switch controller will turn on one or multiple gates to forward the packets, and the selected wavelength channels are multiplexed at the WSS output. The traces of the WDM input packets (shown in black) are illustrated in Figure 9. Each packet labelled with the wavelength channel needs to be switched to the Output 1, while packets labelled with "M" means that multiple gates are enabled for multicasting operation. The switch controls (Ctrl 1 to Ctrl4) generated by the FPGA and applied to the four SOA gates of the WSS are also illustrated in Figure 9. The signals are synchronized with the packets and "on" states correspond to a bias current of 40 mA. The packets of the four channels (CH 1–4) are fast switched (~10 nanoseconds) and the outputs are presented in Figure 9. The traces indicate that the packets are properly switched according to the control signals with a contrast ratio larger than 28 dB.

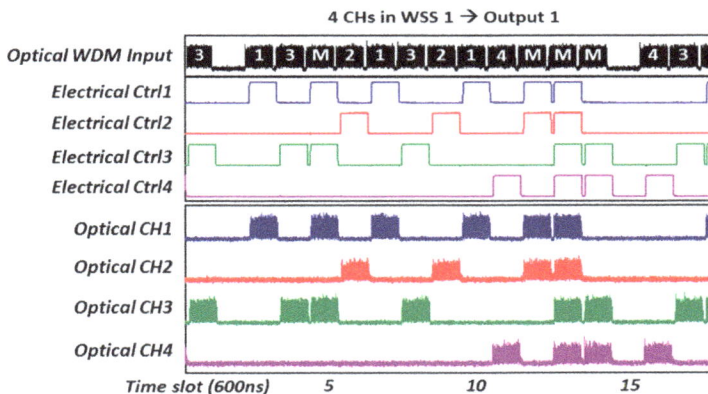

Figure 9. Traces for four channels in WSS1 destined to Output 1.

As a second system assessment, the dynamic switching operation of the four WSSs has been investigated. In this case, packets at wavelength Channel 1 (1525 nm) is switched to one of the four output ports by controlling the Channel 1 SOA gates of the four WSS. The traces of the input packets, the control signals, and the switched outputs are reported in Figure 10. The input packets are labelled with the destination output ports and broadcasting ("B") to two or more ports is also enabled by turning on multiple SOA gates. It can be seen from Figure 10 that the fast dynamic switching operation in space, wavelength, as well as time domains are fully supported by the WDM fast optical switch PIC.

To quantify the performance of the fast optical switch PIC, the BER curves for each WDM channel at different data rates are measured and the results are shown in Figure 11a–c. The back-to-back (B2B) curves are also included for reference. The gate SOA for the channel under test is supplied with a 60 mA driving current and the output is amplified and sent to the BER tester (BERT). When the single wavelength channel is input to the PIC (shown in blue curves), error-free operations with less than

0.5 dB have been measured for Channel 1 (CH 1) and Channel 3 (CH 3) at different data rates while, for Channel 2 (CH 2) and Channel 4 (CH 4), the penalty was around 1 dB at 10 Gb/s, 20 Gb/s, and 2 dB at 40 Gb/s data rate. The eye diagrams of the switched output are also reported and confirm the signal degradation, mainly due to accumulated noise for CH 2 and CH 4. When all four WDM input channels are fed into the switch, the BER results (shown as red curves) indicate a slight performance degradation with an extra penalty of around 0.5 dB for CH 1 and CH 3 and 1 dB for CH 2 and CH 4 compared with single wavelength operation for the different data rates.

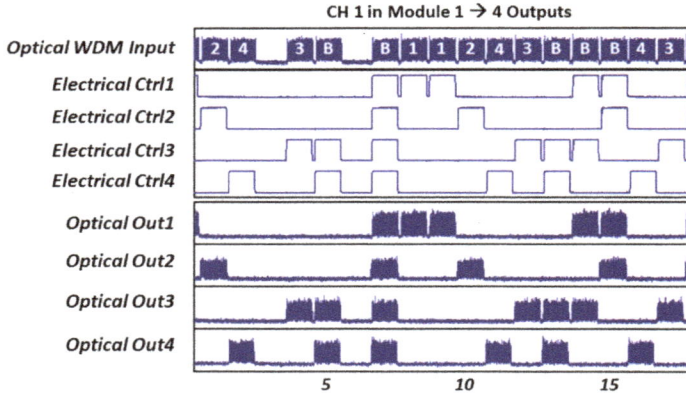

Figure 10. Traces for Channel 1 in Module 1 destined to four outputs.

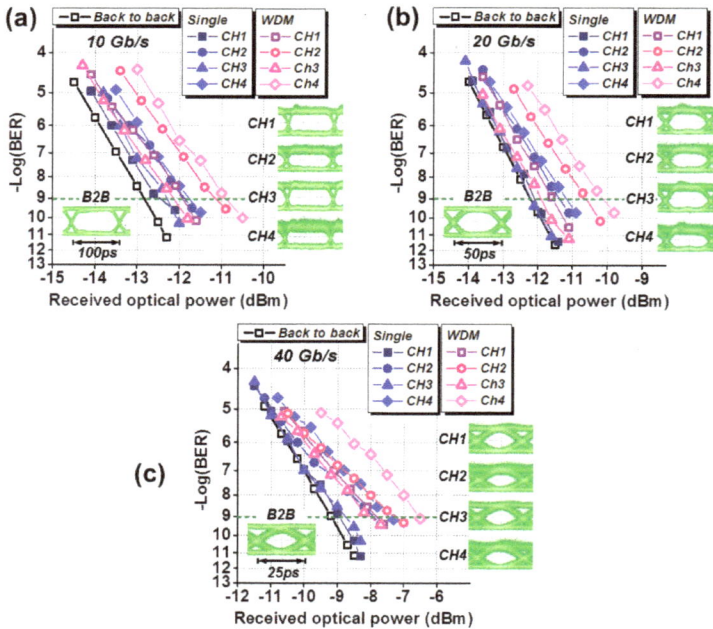

Figure 11. BER curves for single channel and WDM channel input at (**a**) 10 Gb/s; (**b**) 20 Gb/s; and (**c**) 40 Gb/s.

It can be seen from the assessment results that the WDM fast optical switch PIC can dynamically handle the WDM data traffic within a few nanoseconds in space and wavelength domains. Error-free operation has been measured for 10 Gb/s, 20 Gb/s, and 40 Gb/s in single, as well as multiple, WDM channels with <1.5 dB and <3 dB penalty for CH 1/3 and CH 2/4, respectively. As reported in the characterizations, CH 2 and 4 experience extra losses due to the substantial optical power coupling between the channels caused by the not fully resolved waveguides after the de-multiplexer AWG. Therefore, lower input power into the gate SOAs results in OSNR performance degradation, which is also confirmed by the BER curves and eye diagrams in Figure 11. The gain compensation brought by the SOAs and the resulting limited power penalty, indicates the potential scale of the PIC to higher data rates and port counts.

5.2. High-Data and Multi-Level Modulated Traffic

Considering the application of the fast optical switch PIC in DCNs deployed with advanced optical interconnect solutions, the capability of handling the high data-rate and multi-level modulated traffic is a necessity. The potential impact on the signal integrity and OSNR degradation when given lower input power, therefore, should be well addressed. To assess the performance of the fast optical switch PIC in switching the high-capacity and multi-level modulated traffic, the experimental setup shown in Figure 12 is employed. Three types of traffic are generated and tested, namely the 40 Gb/s NRZ-OOK, 20 Gb/s PAM4, and data-rate adaptive DMT. The four WDM channels are de-correlated, amplified, and polarization-controlled before reaching the fast optical switch PIC chip. Module 1, as one of the four identical modules, has been selected for the switching performance assessment. The bias current for the booster SOA and gate SOA are adjusted by an FPGA-based switch controller.

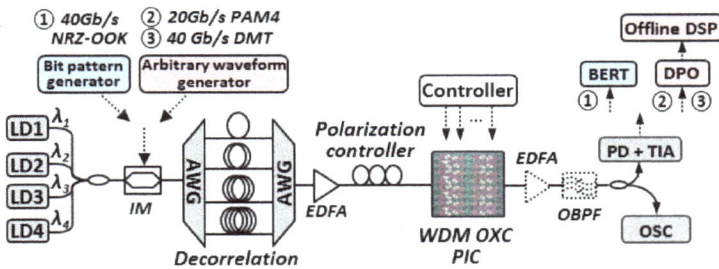

Figure 12. Experimental setup for the performance assessment of NRZ-OOK traffic.

The switching of the packetized 40 Gb/s NRZ-OOK (PRBS 211-1) traffic is first analyzed. The power of each channel launched into the WDM OXC is −4 dBm and the driving current for the booster SOA and gate SOA are 100 mA and 60 mA, respectively. At the output, switched traffic is amplified and sent to the BER tester (BERT). The BER curves, as well as eye diagrams for Channel 1 with a single channel and WDM input applied are shown in Figure 13a. The back-to-back (B2B) curve is also included for reference. Error-free operation with a 0.5 dB power penalty at a BER of 1×10^{-9} has been measured for single channel case. When four WDM channels are fed into the switch, the result indicates slight performance degradation with a penalty of around 1 dB which is mainly due to the noise introduced by the SOAs. The power penalty measured at BER of 1×10^{-9} with different input optical power is plotted in Figure 13b. The bias current for the booster and gate SOAs are varied to equalize the output power of the chip. For single channel input, 14 dB dynamic range is achieved with less than 1.5 power dB penalty. WDM input would cause an expense of 1 dB extra penalty. A smaller dynamic range of 10 dB is obtained for the WDM input case. The required bias current for the booster SOA and gate SOA is also illustrated in Figure 13b. Higher input power would require less driving current for both SOAs to achieve the equalized power output.

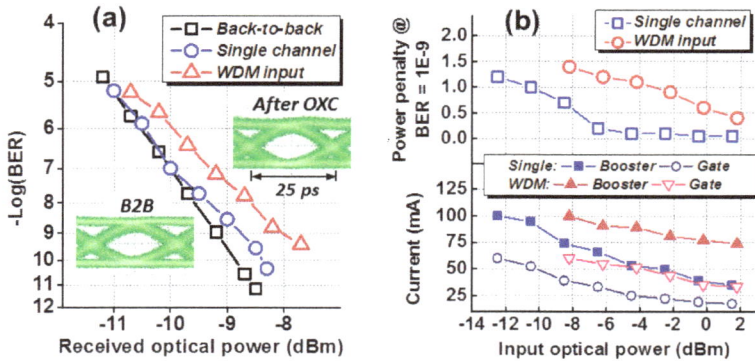

Figure 13. (**a**) BER curves and (**b**) the dynamic range with adjusted SOA current for 40 Gb/s traffic.

As promising interconnect solutions, PAM4 and DMT can effectively boost the link capacity, and the switching performances have been analyzed for the fast optical switch PIC. The WDM channels are modulated with the PAM4 and DMT traffic generated by the 24 GSa/s arbitrary waveform generator. At the switch output, the traffic is received by a real-time 50 GSa/s digital phosphor oscilloscope (DPO) for offline DSP. The BER curves and the eye diagrams of the switched 20 Gb/s PAM4 traffic at Channel 1 are shown in Figure 14a. A power penalty of 0.4 dB and 1 dB at BER of 1×10^{-3} has been observed for the single channel and WDM input, respectively. The optical power dynamic range guaranteeing a BER $< 1 \times 10^{-3}$ is then studied. The power penalty when varying the input power and the corresponding bias current of both SOAs for equalized output are depicted in Figure 14b. Both cases achieve >10 dB dynamic range within a 2 dB power penalty. Similar to the NRZ-OOK traffic, lower input power is limited by the OSNR degradation where the noise is more dominant. A larger penalty is also found for higher input power due to the increased sensitivity to saturation, as confirmed by the eye diagrams in Figure 14b. The situation without the output EDFA is also investigated for the single channel input case, and the dynamic range result is presented in Figure 14b. A similar trend with a smaller power penalty has been observed compared to the presence of EDFA. A smaller dynamic range is mainly due to the inadequate output power level when lower input power is applied.

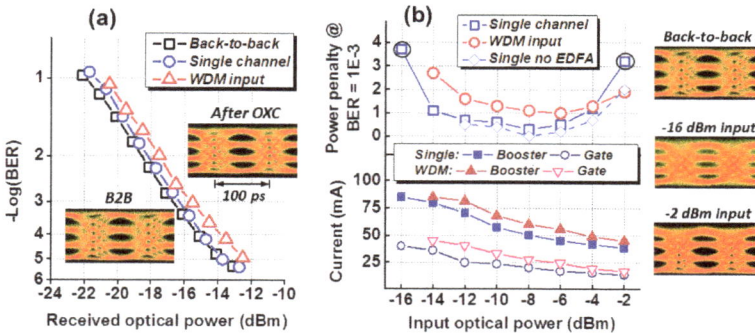

Figure 14. (**a**) BER curves and (**b**) the dynamic range with adjusted SOA current for 20 Gb/s PAM4 traffic.

For the DMT traffic, we first evaluated the effect of the input power to the optical switch PIC chip on the achievable data-rate. As illustrated in Figure 15a, 0 dBm is found to be the optimum (absolute data-rate is not optimized). An example of the optimal bit allocation after bit loading for the 10 GHz DMT with 256 sub-carriers is included in Figure 15b. Channel 1 with 0 dBm optical power is sent to the PIC and the maximum achievable data-rate with an average BER $< 1 \times 10^{-3}$ is reported in Figure 15c. A 1 dB penalty in the data-rate of 32 Gb/s is introduced by the WDM optical switch PIC compared with the B2B traffic.

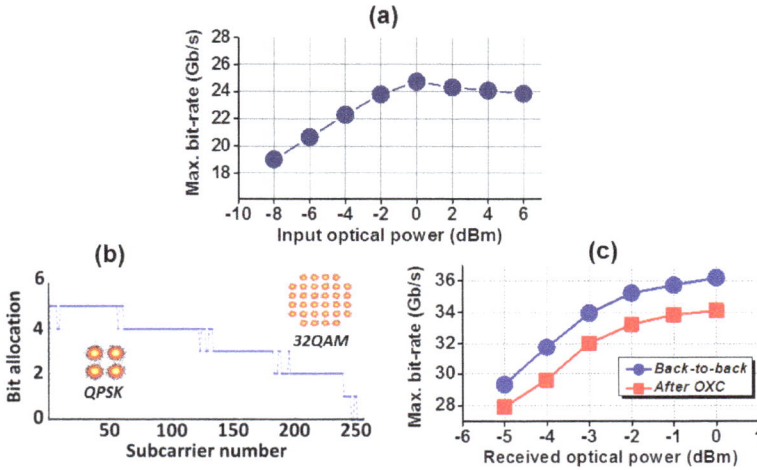

Figure 15. (**a**) Optimal input power; (**b**) bit allocation per subcarrier; and (**c**) maximum data-rate with different received optical powers for DMT traffic.

6. Conclusions

We presented the assessment of a photonic integrated WDM optical cross-connect as one of the main building blocks for interconnecting network elements, as well as computing and storage resources implemented in an optical metro node architecture. The photonic WDM cross-connect node based on semiconductor optical amplifiers (SOA) allows switching data signals in wavelength, space, and time for fully exploiting statistical multiplexing. The assessment of a 4 × 4 photonic WDM cross-connect in terms of switching speed, photonic integrated amplification for loss-less operation, and gain equalization has been reported. The experimental results confirm the potential lossless operation, large optical power dynamic range, low cross-talk <-30 dB, and multicasting operation. Compensation of the losses is a good indication that the modular architecture could scale to a larger number of ports. Moreover, the assessment of the cross-connect switch with multiple WDM channels and diverse modulation formats indicate dynamically switching within a few nanoseconds the WDM data packets and error-free operation with less than a 2 dB power penalty for single channel, as well as WDM input operation, for multiple 10/20/40 Gb/s NRZ-OOK, 20 Gb/s PAM4, and data-rate adaptive DMT traffic.

Considering the practical implementation, due to the identical modular structure, the scalability of the WDM cross-connect switch is mainly determined by possible AWG crosstalk degradation with the increase of the number of ports and by the broadcast and select optical switch. On the AWG crosstalk, it has been demonstrated the potential to have a large (>64) number of ports with very slight degradation of the crosstalk. The splitting loss caused by the broadcasting stage can be compensated by the SOA gates, guaranteeing loss-less operation and sufficient power level for the receiver side, the noise added by the SOA will lead to an OSNR degradation. Therefore, the main limiting factor of scaling out the port-count beyond 64 ports is the OSNR degradation due to the splitting loss experienced by the optical data.

Acknowledgments: The authors would like to thank H2020 METRO-HAUL project (grant number 761727) for supporting this work, and JePPIX for the chip fabrication.

Author Contributions: Nicola Calabretta conceived and designed the PIC WDM cross-connect and wrote the paper. Wang Miao, Ketemaw Mekonnen and Kristif Prifti performed the experiments and analyzed the data with Nicola Calabretta.

Conflicts of Interest: The authors declare no conflict of interest.

References

1. Cisco Systems. *Cisco Global Cloud Index: Forecast and Methodology 2014–2019*; Cisco Systems: San Jose, CA, USA, 2015.
2. CORD Project. Available online: http://opencord.org/ (accessed on 12 June 2017).
3. De Leenheer, M.; Tofigh, T.; Parulkar, G. Open and Programmable Metro Networks. In Proceedings of the Optical Fiber Communication Conference and Exhibition 2016, Anaheim, CA, USA, 20–24 March 2016.
4. Napoli, A.; Bohn, M.; Rafique, D.; Stavdas, A.; Sambo, N.; Poti, L.; Nölle, M.; Fischer, J.K.; Riccardi, E.; Pagano, A. Next generation elastic optical networks: The vision of the European research project IDEALIST. *IEEE Commun. Mag.* **2015**, *53*, 153–162. [CrossRef]
5. Ruffini, M. Metro-Access Network Convergence. In Proceedings of the Optical Fiber Communication Conference and Exhibition 2016, Anaheim, CA, USA, 20–24 March 2016.
6. H2020 METRO-HAUL Project. Available online: https://metro-haul.eu (accessed on 21 June 2017).
7. Ghiasi, A. Large data centers interconnect bottlenecks. *Opt. Express* **2015**, *23*, 2085–2090. [CrossRef] [PubMed]
8. Tucker, R. How to build a petabit-per-second optical router. In Proceedings of the 19th Annual Meeting IEEE Lasers Electro-Optics Society, Montreal, QC, Canada, 29 October–2 November 2006; pp. 486–487.
9. DeCusatis, C. Optical Interconnect Networks for Data Communications. *J. Lightwave Technol.* **2014**, *32*, 544–552. [CrossRef]
10. Kachris, C.; Tomkos, I. A Survey on Optical Interconnects for Data Centers. *IEEE Commun. Surv. Tutor.* **2012**, *14*, 1021–1036. [CrossRef]
11. Farrington, N.; Porter, G.; Sun, P.-C.; Forencich, A.; Ford, J.; Fainman, Y.G.; Papen, G.; Vahdat, A. A demonstration of ultra-low-latency data center optical circuit switching. *Comput. Commun. Rev.* **2012**, *42*, 95–96. [CrossRef]
12. Hemenway, R.; Grzybowski, R.; Minkenberg, C.; Luijten, R. Optical-packet-switched interconnect for supercomputer applications. *J. Opt. Netw.* **2004**, *3*, 900–913. [CrossRef]
13. Liboiron-Ladouceur, O.; Shacham, A.; Small, B.A.; Lee, B.G.; Wang, H.; Lai, C.P.; Biberman, A.; Bergman, K. The data vortex optical packet switched interconnection network. *J. Lightwave Technol.* **2008**, *26*, 1777–1789. [CrossRef]
14. Ye, X.; Yin, Y.; Yoo, S.J.B.; Mejia, P.; Proietti, R.; Akella, V. DOS: A scalable optical switch for datacenters. In Proceedings of the 6th ACM/IEEE Symposium on Architectures for Networking and Communications Systems, La Jolla, CA, USA, 25–26 October 2010.
15. Yan, F.; Miao, W.; Dorren, H.J.S.; Calabretta, N. Novel flat DCN architecture based on optical switches with fast flow control. In Proceedings of the International Conference on Photonics in Switching (PS), Florence, Italy, 22–25 September 2015; pp. 309–311.
16. Miao, W.; Luo, J.; di Lucente, S.; Dorren, H.; Calabretta, N. Novel flat datacenter network architecture based on scalable and flow-controlled optical switch system. *Opt. Express* **2014**, *22*, 2465. [CrossRef] [PubMed]

Review

applied sciences

MDPI

Towards Large-Scale Fast Reprogrammable SOA-Based Photonic Integrated Switch Circuits

Ripalta Stabile

Eindhoven University of Technology, Institute for Photonic Integration (IPI), Department of Electrical Engineering, 5612 AZ Eindhoven, The Netherlands; r.stabile@tue.nl; Tel.: +31-40-247 5387

Received: 14 August 2017; Accepted: 4 September 2017; Published: 7 September 2017

Abstract: Due to the exponentially increasing connectivity and bandwidth demand from the Internet, the most advanced examples of medium-scale fast reconfigurable photonic integrated switch circuits are offered by research carried out for data- and computer-communication applications, where network flexibility at a high speed and high connectivity are provided to suit network demand. Recently we have prototyped optical switching circuits using monolithic integration technology with up to several hundreds of integrated optical components per chip for high connectivity. In this paper, the current status of fast reconfigurable medium-scale indium phosphide (InP) integrated photonic switch matrices based on the use of semiconductor optical amplifier (SOA) gates is reviewed, focusing on broadband and cross-connecting monolithic implementations, granting a connectivity of up to sixteen input ports, sixteen output ports, and sixty-four channels, respectively. The opportunities for increasing connectivity, enabling nanosecond order reconfigurability, and introducing distributed optical power monitoring at the physical layer are highlighted. Complementary architecture based on resonant switching elements on the same material platform are also discussed for power efficient switching. Performance projections related to the physical layer are presented and strategies for improvements are discussed in view of opening a route towards large-scale power efficient fast reprogrammable photonic integrated switching circuits.

Keywords: photonic integrated circuits; SOA; fast optical switching; re-programmability; packet switching

1. Introduction

The global Internet Protocol (IP) traffic is increasing at incredible rates: Cisco projected that global mobile traffic will increase from 3.7 exabytes per month to 30.6 exabytes per month in 2020 [1] and there are no signs that this trend will stop anytime soon [2]. It is not easy to predict implications of this trend in data transmission, but it is clear that Internet traffic is evolving from a relatively steady stream traffic to a more dynamic traffic pattern [3]. New bursty types of packet-based traffic require network flexibility and reconfigurability at a high-speed, as well as high connectivity at low power consumption. This scenario, however, is not sustainable through packet routing and switching in the electronic domain only. Hybrid solutions have been proposed for decades now [4]: photonic hardware is introduced for massive-bandwidth data transport and switching, while routing is handled at more modest GHz rates in the electronic domain.

Planar integrated photonic circuits offer an exciting opportunity to create single chip switching solutions for high capacity packet-compliant data routing [5], since they can potentially reduce the high costs and delays associated with opto-electronic conversion and electronic de/serialization. In the last ten years photonic technologies have quickly matured to the point that an increasing number of components per chip up to few thousands has been demonstrated for function diversity and sophistication [6–9] in the optical communication field where requirements are more severe and urgent.

However, further scalability must be ensured to provide the requested connectivity in future optical switching engines, while maintaining high performance.

In this work, the progress made in electronically actuated monolithic switching circuits based on the use of semiconductor optical amplifiers (SOAs) is reviewed, focusing on circuit-level functionality. After reviewing the integration technologies that have enabled high-connectivity routing in planar optoelectronic circuits in Section 2, we focus on circuits that exploit the indium phosphide (InP) integrated photonics platform. In particular, in Section 3 we address broadband implementations with connectivity of up to sixteen ports in a packet-compliant SOA multi-stage switching matrix. A comprehensive study of the circuit-level gain, loss, and noise performance provides insight into the future scalability of these circuits. In Section 4, the opportunity of scaling connectivity by combining monolithically integrated wavelength- and space-selective elements is described, including demonstrations of packet time-scale reconfigurability. Opportunities for developing a distributed network for intra-chip monitoring and equalization for next generation photonic circuits are discussed in Section 5. In Section 6 we introduce a complementary matrix topology based on resonant elements as developed in the same platform to provide a route for power efficient switching. The strategies for improvements and challenges for future deployment of large-scale fast re-configurable power efficient switch matrices are discussed all over the sections, by looking at figures like connectivity, performance, power consumption, distributed on-chip optical monitoring, and interconnectivity.

2. Monolithically Integrated Fast Optical Switching

Despite the considerable number of concepts and devices proposed and demonstrated over the years, only a relatively small number of integrated switches have scaled to tens of ports. Three-dimensional (3D) microelectromechanical systems (MEMS) have scaled up to hundreds of ports with low (~2 dB) insertion loss and very low power consumption [10]. However, commercially-deployed optical switching engines using MEMS technologies are unwieldy and too slow for future packet-based networking. It is therefore increasingly important to identify scalable optical switching engines which are capable of fast nanosecond-timescale reconfiguration. Photonic integrated circuits can offer an exciting and viable solution for creating a single-chip switching solution for high-capacity data routing. As a notable example, planar integrated MEMS have been deployed in combination with silicon photonics planar circuits to create up to 64 × 64 photonic switch matrices [11]: An electrostatically actuated MEMS waveguide is moved to enable a directional coupling between the input and output buses. These waveguides are low-loss and broadband, with a compact footprint, although the actuation voltages of tens of volts are still relatively challenging and a sub-microsecond switching time is set by the intrinsically limiting mechanical movement.

InP integrated photonics offers high performance amplifiers, switches, modulators, detectors, and de/multiplexers in the same wafer scale process. High speed switch elements have primarily used interferometric and gated switch elements that exploit the InP integrated photonics platform. Interferometers offer the potential for low-current and therefore low-power operation. Considerable work focusing on the use of directional couplers, Y-branch couplers, and Mach-Zehnder interferometers (MZI) has been done. The latter have been configured as 2 × 2 elements in the larger monolithic, multistage 16 × 16 and 32 × 32 switch matrices on LiNbO$_3$ [12–14]. More recently, 8 × 8 switch matrices have been fabricated on Silicon-on-Insulator [15]: Circuit level challenges have focused on improving the modest switch extinction ratios and relaxing the precision requirements for the voltage control. Improved crosstalk and power consumption performance have been achieved using either cascaded Mach-Zehnder interferometers [16] or by including a short semiconductor optical amplifier gate stage [17], to the expenses of a higher number of control electrodes. A different approach is provided by the notable example of the 8 × 8 MOTOR circuit, that integrates eight wavelength converters and one cyclic router, offering the prospect of bandwidth reallocations at the optical layer [18]. However, this class of switch has proved more challenging to integrate.

Semiconductor optical amplifier (SOA) gate switches have been more extensively deployed in system-level studies: Electrical control results are relatively simple and with low voltages; also, tolerance to current variations and excellent crosstalk suppression are typical characteristics [19]. Furthermore, SOA-based gate switches can enable passbands of several Terahertz, colorless and wavelength multiplexed routing [20]. System integrators have demonstrated large optical fabrics using multiple stages and planes of discrete components [21,22]. These successes have led to a renewed interest in large-scale photonic integration as a route to more flexible and lossless routing systems to reduce system level complexity. As a consequence, in the last ten years, photonic integration technology has evolved rapidly to the point that a number of academic and industry research groups are making more and more complex integrated switching circuits. The first lossless 16 × 16 optical switch matrices were demonstrated by Wang et al. [23]. The wide-spread use of amplifying waveguides did, however, compromise the dynamic range and increased power consumption per optical path connections. Subsequent work that we have done using active-passive regrown wafer technologies has allowed further enhancements in terms of optical power handling and noise performance [24]. Moreover, by combining the advantages of high port-count switch fabrics with on-chip wavelength selection [25,26], we have demonstrated even further increases in connectivity of up to 64 inputs wavelength channels on a monolithic integrated prototype [27].

Table 1 summarizes developments in fast SOA-based optical switch matrix technologies for switch sizes of 16 inputs to 16 outputs and above or minimal connectivity of 16 channels. The connectivity is quantified in terms of physical input ports, output ports, and number of unique wavelength specific paths which can be provisioned inside the circuit. These recent developments in switch circuits align approximately to a trend line with a doubling of connectivity in photonics every year [7] and they will be introduced and described in more details in the next sections.

Table 1. Connectivity in fast optical switch matrices for Mach-Zehnder interferometers (MZI) and semiconductor optical amplifier (SOA) gate switch matrices. AWG: arrayed waveguide grating.

	Input Ports × Output Ports	Wavelength Paths	Connectivity	Ref.
SOA-based space switches	16 × 16	1	16	[23]
	16 × 16	1	16	[24]
SOA- and AWG-based cross-connects	4 × 4	4	16	[25]
	4 × 16	4	16	[26]
	8 × 8	8	64	[27]

3. InP SOA-Based Broadband Switch Matrices

Broadband switches are expected to enable considerable energy savings with respect to electronic switching as the line-rate increases, as the data throughput is not directly linked to the actuation energy [28]. So far, the highest connectivity SOA-gate based circuit has been a three-stage 16 × 16 switch with a hybrid Clos-Tree structure [23]. A single active epitaxy was used to enable loss-less on-chip operation and 10 Gb/s routing between selected paths. The use of active waveguides for every component within the chip requires significant levels of current injection to overcome loss, and this in turn leads to a building up in spontaneous emission noise and a compromised optical signal to noise ratio (OSNR) of 14.5 dB/0.1 nm. In [24], we reported the first 16 × 16 switch on a re-grown active-passive epitaxial wafer: The sparse use of SOA gain elements and low loss passive waveguide components is anticipated to allow reduced signal impairment through lower levels of amplified spontaneous emission (ASE) and distortion.

Figure 1a shows the three-stage hybrid Beneš-Tree architecture as implemented with 2 × 2, 4 × 4, and 2 × 2 switches at the input, center, and output stages, respectively. The first and the last stage of SOAs are used as pre-amplifiers and booster amplifiers, respectively. The center stage enables path fast reconfigurability and is replicated in two planes to allow re-arrangeable non-blocking interconnection. The active SOA gates are integrated with the passive broadcast and waveguide routing circuits by using an active-passive epitaxial regrown InP/InGaAsP (Indium Gallium Arsenide Phosphide) wafer. The switch comprises 192,500 μm long SOA gates distributed in six columns and a comparable number of splitters, combiners, and waveguide crossings within a chip area of 4.0 mm × 13.2 mm, for a record number of 480 components on a single photonic chip [7]. To create the photonic waveguide wiring, a four-step inductively coupled plasma (ICP) etch was performed. While the central SOA electrodes are connected to fast programmable drivers for path reconfiguration, the remaining connections at the inputs and at the outputs are made via electrical patch panels to arrays of modular direct current (DC) sources. Both single and multi-path operation are assessed [29]. Polarization controllers are used as the circuit is optimized for transverse electric (TE) polarization.

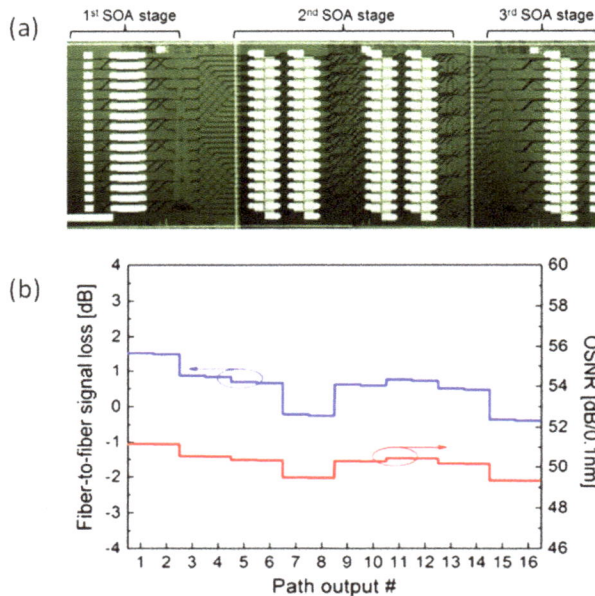

Figure 1. (**a**) A composite photograph of the active-passive indium phosphide (InP) 16 × 16 switch matrix, (**b**) Projected performance of chip (**a**) for a data signal input power of −2.1 dBm and optical signal to noise ratio (OSNR) of 60 dB/0.1 nm for paths from input 1 to output 1-to-16.

Initial characterization is performed to study the component level and path level operation [24]: All quoted on-chip losses come from in-situ measurements. Only thirty-two paths have been tested, due to the chip complexity: the paths from input 1 to output 1 to 32, and paths from input 32 to output 1 to 32—these paths include the shortest and the longest paths of the chip for full loss map understanding. Path yield numbers are reported in the next section for a similar area chip, fabricated on the same wafer. The use of low-loss passive waveguides in combination with high-contrast active gates leads to promising system level metrics in terms of optical signal to noise ratio with a best case value of 28 dB/0.1 nm. However, due to the moderate scale of integration, the space switch scheme presents high path losses: The fiber-to-fiber signal losses increase as path length and complexity increase, because of the additional traversed components. The measured total passive losses along the thirty-two paths include an inherent 30 dB loss contribution resulting from the broadcast-select capability and an excess 37 dB loss resulting from component-level imperfections, with a maximum loss variation of 12.8 dB. The gain from the SOA gates can compensate much of the excess component losses, while the inherent losses of about 30 dB coming from the broadcast-enabled architecture are present in any given path. The circuit-level losses were analyzed by evaluating the losses in comparable circuits and through direct measurement. The noise performance was predicted by estimating the optical power map for the signal and noise within the switch matrix circuit and including the additional amplified spontaneous emission noise contributed at each SOA gate. Each of the three stages of the SOA gates in the circuit was expected to operate with a mean gain of 12.9 dB and a mean amplified spontaneous emission power density of -47.3 dBm/0.1 nm. The operating optical input power for the input SOA was -11.6 dBm and for the central and output SOAs was -21 dBm, so the 500-µm-long gates were operated far from the optical saturation power. The electrical bias conditions were 75 mA and 1.5 V. The agreement for the implemented elements is excellent, giving confidence that the loss and noise performance of the circuits can be clearly attributed to the excess losses of the components. The excess losses can, however, be significantly reduced by using optimized components like mode-size adaption at the facets [30], low-parasitic-reflection splitters [31], low-loss waveguide crossing through a broadening at the intersecting waveguides [32], and low waveguide losses of the order of 0.5 dB/cm for minimized p-doping level in the passive waveguides [33]. The projected performance for the 16×16 switch matrix with optimized component losses are shown in Figure 1b: lossless circuit level operation is enabled and a radical impact on the optical signal to noise ratio is found by being increased up to about 50 dB/0.1 nm. This would correspond to a circuit level noise figure of 13.7 dB, of which 5.3 dB is attributable to losses between the input side fiber and the first SOA gate. This analysis suggests how a further increase in connectivity is possible, but requires careful design of the chain of cascaded SOAs in a multistage optical switch: reduced losses at the input of the first SOA stage can provide improved switch noise figures. Also, the losses per stage must be designed to be balanced by the SOA-gain stage, keeping the SOA far from the saturation point, in a chain of multiple optical amplifiers.

The operating current and voltage per path are currently 240 mA and 1.5 V for the three gates. If an on-chip preamplifier is included for on-chip gain, the operating current (power) is expected to increase to 340 mA (0.55 W/path). A few watts of power consumption is then expected for the fully loaded chip. The temperature is kept at 15 °C by fixing the chip on a temperature-stabilized water-cooled copper mount. When compared with an all-active approach, the active-passive integration scheme enables order-of-magnitude noise performance improvement as well as reduced energy requirement.

This analysis opens a route to broadband switches with connectivity higher than sixteen IOs. Larger level connectivity clearly implies a larger area that is, however, limited not only by the maximum available InP wafer size, but also by the difficulty of a uniform wafer processing. In Section 7 the author comments on complementary approaches to pursue large-scale integration.

4. Monolithically Integrated InP 8 × 8 Space and Wavelength Selecting Switch

Orders of magnitude increases in reconfigurability become possible when the use of wavelength division multiplexing (WDM) is combined with on-chip wavelength selective routing: The 8×8 integrated cross-connect represents an elegant example of an in-plane circuit which provides a connectivity of up to 64 input and output channels [27]. Here, we perform both space domain switching and wavelength domain channel selection within the same monolithic circuit in order to show the highest level of connectivity and data capacity on a single InP photonic integrated chip. The architecture for broadband photonic and wavelength selective cross-connection is shown in Figure 2a. The circuit consists of a combination of a broadband photonic switch stage (PSS) for port selection and a wavelength selecting stage (WSS) for color selection, both based on the use of SOAs. Eight colorless WDM input ports connect to an array of pre-amplifying SOAs. Broadcast shuffle networks map each of the inputs to the eight output ports. The first stage of SOA gates is used for input port selection and connects to eight cyclic arrayed waveguide gratings (AWGs). The AWGs route the different wavelength channels to the second stage array of broadband SOAs to perform wavelength selection. Selected channels are aggregated with broadband fan-ins to provide the eight colorless outputs. The grey-scale layers in Figure 2a are the other seven identical layers implemented in the integrated circuit for both stages. The red boxes are the SOA-gates used in each of the stages.

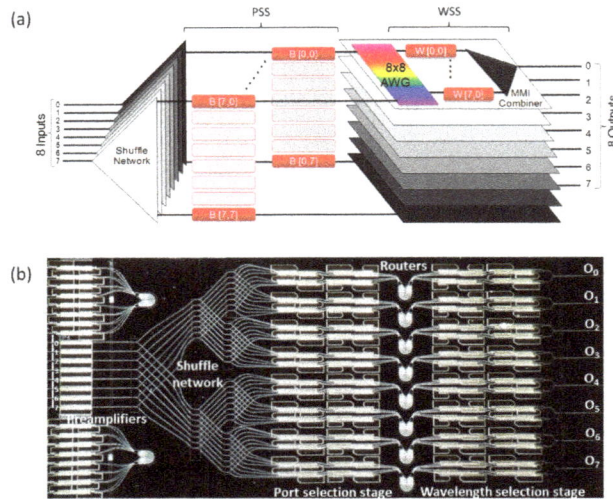

Figure 2. (**a**) Architecture and (**b**) composite image of the active-passive InP 8 × 8 cross-connect.

The device is realized on a re-grown active-passive InGaAsP/InP epitaxy. An optical image of the chip is shown in Figure 2b. The eight pre-amplified colorless inputs are broadcast through a shuffle network made of cascaded 1 × 2 multimode interferometer (MMI) splitters. After port and color selection, the selected channels are combined with broadcast cascaded 2 × 1 MMI couplers. The active islands in the InP wafer are used for SOA gates and preamplifiers, while the passive regions are used for waveguide wiring, splitters, and cyclic routers. The cyclic AWG used in the WSS is designed with a channel spacing of 3.2 nm (400 GHz) and a free spectral range of 25.6 nm. All input and output waveguides are positioned on a 250 μm pitch to enable simultaneous access to all the ports using a commercially sourced lensed fiber array. The 136 SOA contacts are wire-bonded to an electronic printed circuit board (not shown). The chip is attached with conductive epoxy to a water-cooled block. The total footprint of the switch is 16.4 × 6.7 mm^2. Electrical I-V characteristics are analyzed to identify 126 SOAs with good diode characteristics out of a total of 136. Two fails are attributable to an imperfect lithographic step causing electrode detachment, six are associated with bond-wire detachments, and two are short circuits. From the photonic switching stage to the output side, 432 paths out of 512 paths connections are verified: 84% of the paths from the input to the output side of the chip are electrically and therefore optically connected in this first prototype. Each path includes three SOAs as in the 16 × 16 broadcast switch matrix, but the losses are modest, which indicates a more straightforward route to further scaling. However, in this case the multi-stage switch is characterized by an asymmetric chain of SOAs: Moving to a 16 × 16 cross-connect, for example, implies the introduction of six splitters after the first amplifier which needs to be compensated by a longer second-stage SOA.

Multi-path WDM 10 Gb/s data routing is demonstrated through the experimental control plane schematically shown in Figure 3. Very importantly, a common reference clock generator is needed for the programmable logic and the bit error rate test equipment to allow synchronization between the routed data and the switch controller. At the optical plane, four different wavelength channels, λ_0, λ_1, λ_2, and λ_3, with a nominal channel separation of 400 GHz are multiplexed and modulated using a Mach-Zehnder modulator. The WDM signal is then amplified, de-multiplexed, de-correlated, and used as input to the device with 0 dBm channel power. The chip output is connected to a pre-amplified optical receiver after a 0.95 nm bandwidth filter for broadband noise rejection. An average total current per path of the order of 140 mA is used. Optical signal to noise ratios greater than 27.0 dB for 0.1 nm resolution bandwidth and a mean 13.3 dB on-chip loss are measured for each of the channels of a multiplexed channel, as evaluated at a representative path from input 0 to output 0 [27]. With the maximum number of three SOAs on a single path, no oscillation is observed. When moving to a bigger number of gates, the inclusion of isolation sections may be needed.

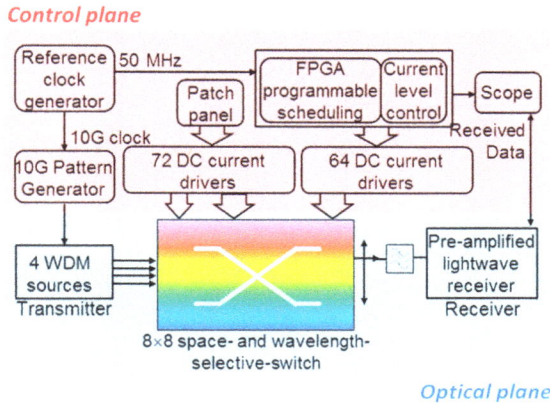

Figure 3. Setup for wavelength division multiplexing (WDM) single and multi-port simultaneous routing. FPGA: field programmable gate array.

The electrical plane comprises fast current drivers and programmable logic. A multiple current source provides DC currents to the pre-amplifying SOAs. Fast current drivers connect to the SOA selector gates of the integrated cross-connect to enable path and channel selection. The current levels for each of the SOA selector gates is separately programmed by means of digital-to-analog converters (DACs) and a central microcontroller. An Altera Stratix III field programmable gate array (FPGA) provides time-slotted control signals to the nanosecond rise-time current drivers for optical path selection. Bit-level synchronization is achieved between the routed data and the switch controller by means of a common reference clock generator between the FPGA and the bit error rate (BER) test equipment.

To demonstrate simultaneous WDM multi-path data routing the wavelength multiplexed data is split into four copies using broadband splitters and launched into four input ports: 0, 1, 6, and 7 of the chip. Sixteen channels are routed within the 8 × 8 cross-connect to output port 0. All four input paths are sequentially enabled with a round-robin scheduling and loaded with WDM input signals for dynamic multi-path WDM reconfigurability studies. The correspondent SOAs are driven by periodic 1 μs pulses with 60 ns guard-bands programmed via the FPGA. Figure 4a shows the time traces of the output WDM signals. Along the x-axis the selected input port changes at each time slot as for the round-robin scheduling. The four sets of wavelengths are selected using external optical filters and are displayed as four separate graphs. The time traces appear to be clean and well resolved. The optical power is levelled for each wavelength to within 2 dB, as the current values have been optimized for equalized output power from time-slot to time-slot. Figure 4b also shows the fall and rise time for the output signal when moving from one time slot to the next one: The switching occurs within 5 ns [34]. Power penalty is expected to be dominated by amplified spontaneous emission from the transmitter side amplifier and the cross-connect. For higher input powers and number of wavelengths per channels, new optical amplifier concepts based on low-overlap between the mode and the gain area may prevent unwanted cross-gain modulation.

The requirements for reduced energy use also require the photonic components to be abstracted to enable tractable network level implementations: Multicast capability, packet-time-scale reconfigurability, power levelling irrespective of signal path, distributed signal quality monitoring, and data rate transparency have been implemented at the chip level in order to enable an abstracted digital control plane [34–36]. The first investigation on the impact of varied switch loads from unicast to multicast to broadcast on device performance was also performed for this chip [35]. Electrical power dissipation may be estimated from the mean bias current (50 mA) and the mean voltage across the SOAs (1.2 V). The preamplifiers and wavelength-select stage SOAs are operated continuously and the space-select stages are scheduled. This leads to an electrical input power of 0.78 W for unicast and 2.88 W for broadcast operation. If dynamic scheduling was implemented for the wavelength-select stages as well, a more modest 1.44 W electrical power would be required with only 1 dB gain reduction. Further energy savings are also feasible through optical circuit level optimizations [34].

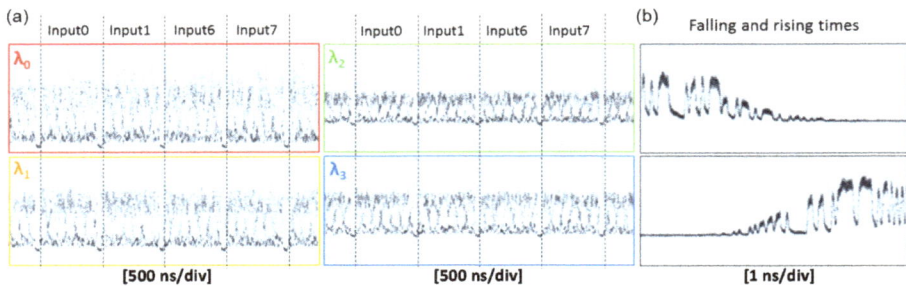

Figure 4. Time traces (**a**) rise and fall time (**b**) for dynamic multi-port WDM routing at output 0.

5. On-Chip Distributed Power Monitoring

The recent demonstrations of multi-path, multi-wavelength packet scheduling highlight the potential for highly reconfigurable networking with relatively simple control planes, but this places stringent demands on the channel monitoring. So far signal monitoring is being performed at the edge of the network, and is not well adapted for an optical packet paradigm. Distributed optical performance monitoring (OPM) is the key tool for future optical networking. In particular, the optical signal-to-noise ratio (OSNR) monitoring can provide information on the intra-chip signal transmission quality. Intra-chip OSNR measurement has been demonstrated through both discrete [37] and integrated [38–40] solutions, but has never been integrated within a switching circuit. However, the most advanced circuits do already contain the essential components. A combination of SOA gate and AWG technology can provide on-chip integration of active elements and filtering functionality for a more straightforward route to signal and noise detection.

We have recently proposed and demonstrated an OSNR meter co-integrated with the 8×8 cross-connect, presented in Section 4, using switch gates as photo-detectors (PDs), to enable the analysis of signal quality in an integrated packet routing engine [36]. An electronic pre-calibration scheme is proposed which can be performed before chips are selected for packaging and is done to measure the filtered integrated optical power after the AWGs, which is then used to calculate a noise power spectral density correction factor to be used for the intra-chip OSNR calculation. For the in-service operation, the pre-amplifier is used to compensate fiber-chip coupling of the in-coming signal, and the wavelength-select gates are used again in detector mode to separately measure signal and noise power. Figure 5a shows the integrated currents measured at all the PDs. The signal level increases exponentially with input power. At low input powers of -30 dBm power, the measured on-chip signal power approaches the integrated in-band noise. This is expected to be dominated by the space-select ASE.

The OSNR is calculated as the ratio between the total integrated current received at the PD6, which is corrected for in-band noise and the appropriately scaled total out-of-band noise contribution. This is plotted as a function of the input power in Figure 5b. The red curve shows the OSNR versus in-fiber optical input powers: An OSNR dynamic range extending from 6 to 40 dB/0.1 nm is demonstrated. The calculated noise figure of the last SOA gate is also plotted as the difference between the calculated on-chip OSNR and the measured off-chip OSNR.

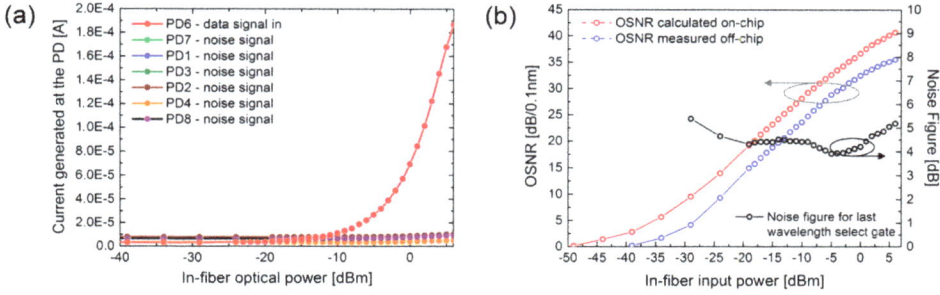

Figure 5. (**a**) Integrated photocurrent measurements (**b**) On-chip OSNR calculation (red curve) and off-chip OSNR measurement. The noise figure of the SOA gate is also plotted (black curve).

The possibility of providing intra-chip measurement of signal OSNR opens up the opportunity to provide real-time equalization per channel in WDM optical switches: on-chip OSNR and power monitoring per channel increase network awareness at the physical layer and allow for power channel equalization. Concepts of WDM space switches followed by co-integrated channel separation for per-channel monitoring and equalization can become feasible at the packet level. The schematic of the foreseen concept for an aware space switch is displayed in Figure 6a.

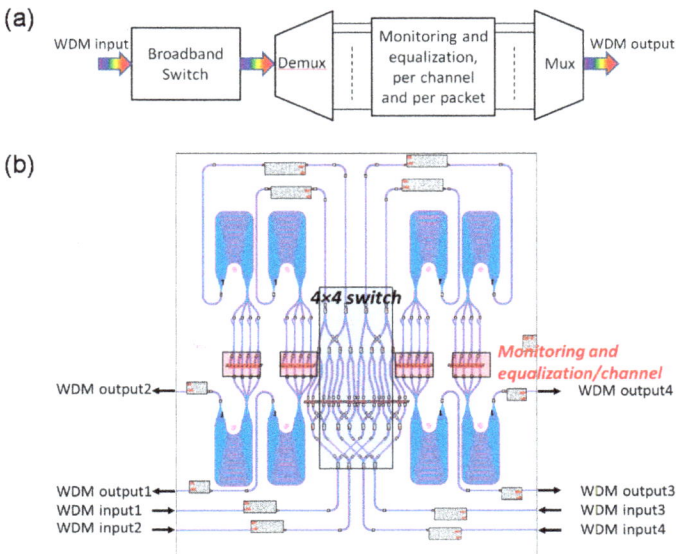

Figure 6. (**a**) Concept of a co-integrated per-channel monitoring and per channel-equalization block for broadband photonic switches and (**b**) tentative photonic integrated circuit design.

A tentative design for an aware WDM 4 × 4 space switch is shown in Figure 6b: the light blue box indicates the 4 × 4 space switch and the four pairs of AWGs are for each of the multiplexed channels. The red boxes include the power monitoring, implemented per channel as an integrated 5:95 tap based on a directional coupler design, and an in-line 3 dB SOA-based attenuator, for power adjustment at the chip output. This chip, fabricated on the generic InP technology, takes two die cells of space for a total size of 4.6×8 mm^2.

6. InP MRR-Based Switching Cross-Point Matrices

Micro-ring resonator (MRR) elements have been considered extensively, given the increased availability of silicon on insulator technology for photonic circuits. Higher-order ring resonators offer a flattened-passband response that enables wavelength-tolerant, broadband routing in a cross-grid array [41]. Faster actuation has proven more challenging with cross-point architectures generally because the physical size of the switching elements has largely precluded the use of fast electro-optic switching. One potential exception is the vertically coupled SOA switch [42]. Recently, we have explored the use of photonic integrated circuits based on ring-resonators on the InP photonics platform. By using InP integrated phase shifters for phase matching and switching activation, these switch matrices provide fast switching and potential low electrical power consumption. Prior work in InP photonics has focused on using first-order micro-ring resonators for this purpose [43,44], but higher-order resonators have been shown to decouple bandwidth and the signal extinction implying that both parameters can be optimized individually. As an added incentive, the enhanced bandwidth of such devices also increases their tolerance to variations in fabrication [41].

Recently, we have designed and fabricated the first 4 × 4 InP switch matrix based on third-order resonators [45,46]: three race-track-like ring resonators are coupled through directional couplers (Figure 7a). The local refractive index changes due to the electric field and the carrier depletion of a reverse biased junction is used to tune the resonant-based switch elements and represents an attractive power-efficient alternative. Figure 7b indicates the phase modulator placement. Fast electro-optic tuning at close-spaced phase modulators is used to allow data to drop to the next port or pass through. The matrix is made of sixteen switching elements which are based on these third-order ring resonators. The full mask layout and a photograph of the realized entire switch matrix are shown in Figure 7c,d, respectively.

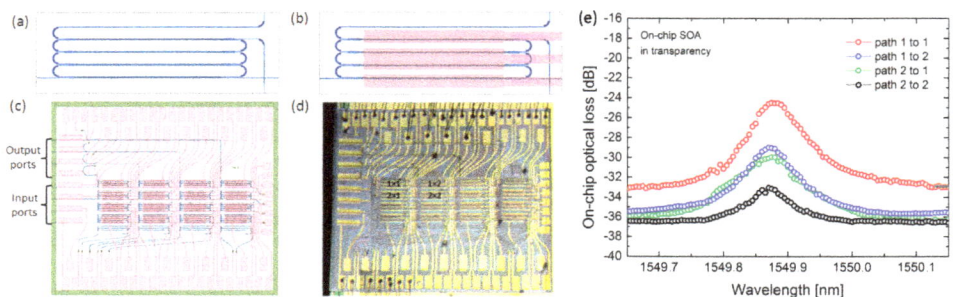

Figure 7. (**a**) Single switching element composed of three race-track micro-rings; (**b**) Electrodes (in red) placed on the top of directional coupler for fine tuning; (**c**) Full layout of the chip, including the optical wiring layer (in black) and the metal wiring (in red); (**d**) Photograph of the fabricated chip; (**e**) Spectral responses of four measured paths: 1×1, 1×2, 2×1, and 2×2.

The desired coupling coefficients are determined by varying the gap separating the directional couplers, while keeping fixed the coupling lengths. The total chip size is about 4.5×3.8 mm^2 and includes 500 μm long SOAs placed after each input-output port to compensate for coupling losses and for power equalization at the optical outputs. Figure 7e shows the measured spectra for all the four paths. The total 12 dB (2 × 6 dB) fiber-to-chip coupling losses are already subtracted from the plots. All four symbol curves are centered at 1549.87 nm and a 3 dB pass-bandwidth of 11.25 GHz (0.09 nm) is measured for each of the switch elements, which is close to the nominal value. The stop-band part of the higher loss spectra is believed to be limited by the SOA noise. On-state on-chip losses for the single switching element are of the order of 6 dB. To enable high connectivity switching fabrics, smaller and lower-loss switch elements will become important. Techniques to further reduce the on-chip losses are currently being investigated: the use of high resolution lithography is being explored to enable higher density and lower on-chip losses. Moreover, the combination of high precision lithography for shorter couplers, higher efficiency quantum well phase modulators, and the use of all-deep ring resonators, which avoid the need for mode converters, is believed to enable further scaling up to the same order as for SOI ring-resonators based cross-point matrices [45].

Ten and twenty billion bits per second (10 Gbps and 20 Gbps) data routing is performed across a combination of four different paths in the circuit with error-free operation with a maximum power penalty of 2.6 dB [47]. The on-state electrical power consumption is recorded to be only 79 μW for the single switch element, which leads to a minimum energy consumption of 8 fJ/s for 10 Gb/s operation, identifying a potential route for energy efficient optical switching. Fast switching of a few nanoseconds for the rise and the falling time is also verified.

7. Discussion

The projected performance of recently realized medium-scale InP photonic fast switching circuits have been presented. This helps in defining the strategies towards large-scale InP photonic integrated circuits (PICs) switches in relation to figures like connectivity, physical layer performance, power consumption, distributed on-chip monitoring, and interconnectivity.

High-radix 4 × 4 SOA-based switch elements have been efficiently configured in multistage networks to enable high connectivity with a limited number of stages, including 2 × 2 SOA-based switches at the input and at the output. Three switch stages provide a good compromise between optical wiring complexity, signal integrity, and control complexity. By using active–passive multi-stage SOA monolithic integration we have enabled switching matrices with up to 480 components, connecting 16 inputs to 16 outputs [24] (Figure 1a). This can potentially route massive bandwidth signals independently of signal wavelength and in a number of multiplexed wavelength channels. We have demonstrated that orders of magnitude increases in reconfigurability is possible when the use of wavelength division multiplexing is combined with on-chip wavelength selective routing: The 8 × 8 InP integrated cross-connect provides a connectivity of up to 64 input and output channels [27] (Figure 2a). Further increases in connectivity are possible but require careful design of the chain of cascaded SOAs and losses per stage. Also, reduced losses at the input of the first SOA stage must be provided for low signal degradation.

Complementary approaches for achieving large-scale integrated optical switch matrices require a complex mesh of interconnection which is not well supported by single layer waveguides as conceived so far. Further scaling motivates more sophisticated optical wiring. Two strategies for achieving higher connectivity are here highlighted. The InP-Membrane-On-Silicon (IMOS) platform is a photonic integration researching InP-based membranes on top of silicon chips [48]. This approach represents a valid example of how to shrink size, while retaining gain and enabling future integration on complementary metal-oxide semiconductor (CMOS) circuitry, through a low thermal conductivity polymer layer. Also, monolithic multi-layer InP integration can be used to enable three-dimensional connectivity for large-scale photonic integrated circuits, potentially offering a step change in optical circuit connectivity and removal of crosstalk at the top-bottom waveguide intersections [49,50].

As the data throughput is not directly linked to the actuation energy, such broadband switches are expected to enable considerable energy savings as the line-rate increases [28]. The involved electrical input energy for chip reconfiguration is of the order of a few watts [24], but further electrical

Appl. Sci. **2017**, *7*, 920

power reduction is desirable. New architectures and technologies must be conceived that provide packet-scale switching functionality while being power efficient. The combination of SOA technology and InP electro-optic effect offers the opportunity of achieving potentially low power consumption and fast switching in photonic integrated circuits based on ring resonators: the first InP third-order ring resonator-based switch matrix fabricated on active-passive integrated InP platform [45] is characterized by an on-state electrical power consumption of only tens of µW per switch element, which is four order of magnitude lower than in a conventional forward bias SOA-gate. The worst case 6 dB losses between the paths are expected to be dominated by the off-state losses which may be reduced with further building block optimization. The low extinction transfer function might be due to obtained off-target coupling coefficients at the directional couplers, which we believe we can overtake by patterning the circuit through deep- ultraviolet (UV) lithography: higher control of critical dimension losses and reduced sizes switch element are expected to enable lossless operation of cross-point matrices.

The recent demonstrations of multi-path, multi-wavelength packet scheduling highlight the potential for highly reconfigurable networking with relatively simple control planes, but this places stringent demands on the channel monitoring: Distributed optical performance monitoring (OPM) is the key tool for future optical networking. The recently proposed and demonstrated OSNR meter, co-integrated with a photonic integrated switch using switch gates as detectors [36], opens the route to on-chip networks where it is possible to "sense and act" in order to maintain the signal integrity or to diverge data through a different path.

8. Conclusions

InP integrated photonic circuits offer a powerful route for combining advanced routing and signal-processing functions onto one single chip. Multicast capability, packet time-scale reconfigurability, power levelling irrespective of signal path, distributed signal quality monitoring, and data rate transparency can be implemented at the chip level, which will enable an abstracted digital control plane. Achieving this level of signal processing and component density may have impact beyond the immediate pressure points in data communications and find potential new applications from sensor readout technologies to optical beam-forming technologies and optical computing.

Acknowledgments: Kevin Williams is acknowledged for fruitful and inspiring discussions. The work has been supported by the Dutch technology foundation STW.

Conflicts of Interest: The author declares no conflict of interest.

References

1. Cisco Visual Networking Index: Global Mobile Data Traffic Forecast Update, 2015–2020 White Paper. Available online: https://www.cisco.com/c/en/us/solutions/collateral/service-provider/visual-networking-index-vni/mobile-white-paper-c11-520862.html (accessed on 28 March 2017).
2. Scaling the 'Data Tsunami' and Market Opportunity for Optical Interconnect In Data Centers. Available online: https://www.lightcounting.com/News_030613.cfm (accessed on 6 March 2013).
3. The Zettabyte Era: Trends and Analysis. Available online: https://files.ifi.uzh.ch/hilty/t/Literature_by_RQs/RQ%20102/2015_Cisco_Zettabyte_Era.pdf (accessed on 7 June 2017).
4. Williams, K.A.; Stabile, R. Advances in integrated photonic circuits for packet-switched interconnection. *Proc. SPIE* **2014**. [CrossRef]
5. Doerr, C. Planar lightwave devices for WDM. In *Optical Fiber Telecommunications IV, A-Components*; Kaminov, I.P., Li, T., Eds.; Bell Laboratories Lucent Technologies: Homdel, NJ, USA, 2002; pp. 402–477.
6. Liu, W.; Li, M.; Guzzon, R.S.; Nrberg, E.J.; Parker, J.S.; Lu, M.; Coldren, L.A.; Yao, J. A fully reconfigurable photonic integrated signal processor. *Nat. Photonics* **2016**, *10*, 190–196. [CrossRef]
7. Smit, M.K.; Leijtens, X.; Ambrosius, H.; Bente, E.; Van der Tol, J.; Smalbrugge, B.; De Vries, T.; Geluk, E.J.; Bolk, J.; Van Veldhoven, R.; et al. An introduction to InP-based generic integration technology. *Semicond. Sci. Technol.* **2014**, *29*, 1–41. [CrossRef]

8. Kish, F.; Nagarajan, R.; Welch, D.; Evans, P.; Rossi, J.; Pleumeekers, J.; Dentai, A.; Kato, M.; Corzine, S.; Muthiah, R.; et al. From visible light-emitting diodes to large-scale III–V photonic integrated circuits. *Proc. IEEE* **2013**, *101*, 2255–2270. [CrossRef]

9. Summers, J.; Vallaitis, T.; Evans, P.; Ziari, M.; Studenkov, P.; Fisher, M.; Sena, J.; James, A.; Corzine, S.; Pavinski, D.; et al. 40 Channels × 57 Gb/s monolithically integrated InP-based coherent photonic transmitter. In Proceedings of the European Conference on Optical Communication (ECOC), Cannes, France, 21–25 September 2014.

10. Madamopoulos, N.; Kaman, V.; Yuan, S.; Jerphagnon, O.; Helkey, R.J.; Bowers, J.E. Applications of large-scale optical 3D-MEMS switches in fiber-based broadband-access networks, Photon. *Netw. Commun.* **2010**, *19*, 62.

11. Seok, T.J.; Quack, N.; Han, S.; Zhang, W.; Muller, R.S.; Wu, M.C. 64 × 64 Low-Loss and Broadband Digital Silicon Photonic MEMS Switches. In Proceedings of the 2015 European Conference on Optical Communication (ECOC), Valencia, Spain, 27 September–1 October 2015.

12. Duthie, P.J.; Shaw, N.; Wale, M.; Bennion, I. Guided wave switch array using electro-optic and carrier depletion effects in Indium Phosphide. *Electron. Lett.* **1991**, *27*, 1747–1748. [CrossRef]

13. Okayama, H.; Kawahara, M. Prototype 32 × 32 optical switch matrix. *Electron. Lett.* **1994**, *30*, 1128–1129. [CrossRef]

14. Murphy, E.; Murphy, T.O.; Ambrose, A.F.; Irvin, R.W.; Lee, B.H.; Peng, P.; Richards, G.W.; Yorinks, A. 16 × 16 nonblocking guided-wave optical switching system. *J. Lightwave Technol.* **1996**, *14*, 352–358. [CrossRef]

15. Lee, B.G.; Rylyakov, A.; Green, W.M.J.; Assefa, S.; Baks, C.W.; Rimolo-Donadio, R.; Kuchta, D.; Khater, M.; Barwicz, T.; Reinholm, C.; et al. Four and eight port photonic switches monolithically integrated with digital CMOS logic and driver circuits. In Proceedings of the Optical Fiber Communication Conference and Exposition and the National Fiber Optic Engineers Conference (OFC/NFOEC), Anaheim, CA, USA, 17–21 March 2013.

16. Van Campenhout, J.; Green, W.M.; Assefa, S.; Vlasov, Y.A. Drive-noise-tolerant broadband silicon electro-optic switch. *Opt. Exp.* **1999**, *19*, 11568–11577. [CrossRef] [PubMed]

17. Cheng, Q.; Wonfor, A.; Wei, J.L.; Penty, R.V.; White, I.H. White, Low-Energy, High-Performance Lossless 8 × 8 SOA Switch'. In Proceedings of the Optical Fiber Communication Conference (OFC), Los Angeles, CA, USA, 22–26 March 2015.

18. Nicholes, S.C.; Masanovic, M.L.; Jevremovic, B.; Lively, E.; Coldren, L.A.; Blumenthal, D.J. An 8 × 8 InP monolithic tunable optical router (MOTOR) packet forwarding chip. *J. Lightwave Technol.* **2010**, *2*, 641–650. [CrossRef]

19. Tanaka, S.; Jeong, S.H.; Yamazaki, S.; Uetake, A.; Tomabechi, S.; Ekawa, M.; Morito, K. Monolithically integrated 8:1 SOA gate switch with large extinction ratio and wide input power dynamic range. *J. Quantum Electron.* **2009**, *45*, 1155. [CrossRef]

20. Stabile, R.; Albores-Mejia, A.; Rohit, A.; Williams, K.A. Integrated optical switch matrices for packet data networks. *Nat. Microsyst. Nanoeng.* **2016**, *2*, 15042. [CrossRef]

21. Shacham, A.; Small, B.A.; Liboiron-Ladouceur, O.; Bergman, K. A fully implemented 12 × 12 data vortex optical packet switching interconnection network. *J. Lightwave Technol.* **2005**, *23*, 3066–3075. [CrossRef]

22. Luijten, R.; Grzybowski, R. The osmosis optical packet switch for supercomputers. In Proceedings of the Optical Fiber Communication Conference, San Diego, CA, USA, 22–26 March 2009.

23. Wang, H.; Wonfor, A.; Williams, K.A.; Penty, R.V.; White, I.H. Demonstration of a lossless monolithic 16 × 16 QW SOA switch, Post-deadline paper. In Proceedings of the 35th European Conference on Optical Communication (ECOC '09 (ECOC)), Vienna, Austria, 20–24 September 2009; pp. 1–2.

24. Stabile, R.; Albores-Mejia, A.; Williams, K.A. Monolithic active-passive 16 × 16 optoelectronic switch. *Opt. Lett.* **2012**, *37*, 4666–4668. [CrossRef] [PubMed]

25. Rohit, A.; Bolk, J.; Leijtens, X.J.M.; Williams, K.A. Monolithic nanosecond-reconfigurable 4 × 4 space and wavelength selective cross-connect. *J. Lightwave Technol.* **2012**, *30*, 2913–2921. [CrossRef]

26. Calabretta, N.; Dorren, H.; Williams, K.A. Monolithically integrated WDM cross-connect switch for nanoseconds wavelength, space, and time switching. In Proceedings of the European Conference on Optical Communication (ECOC), Valencia, Spain, 27 September–1 October 2015; pp. 1–3.

27. Stabile, R.; Rohit, A.; Williams, K.A. Monolithically integrated 8 × 8 space and wavelength selective cross-connect. *J. Lightwave Technol.* **2013**, *32*, 201–207. [CrossRef]

28. Albores-Mejia, A.; Gomez-Agis, F.; Dorren, H.J.S.; Leijtens, X.J.M.; Smit, M.K.; Robbins, D.J.; Williams, K.A. 320 Gb/s data routing in a monolithic multistage semiconductor optical amplifier switching circuits. In Proceedings of the 36th European Conference and Exhibition on Optical Communication (ECOC), Torino, Italy, 19–23 September 2010.

29. Stabile, R.; Albores-Meija, A.; Williams, K.A. Dynamic multi-path routing in a monolithic active passive 16 × 16 optoelectronic switch. In Proceedings of the Optical Fiber Communication Conference and Exposition (OFC) National Fiber Optic Engineers Conference (NFOEC), Anaheim, CA, USA, 17–21 March 2013.

30. Langley, L.N.; Robbins, D.J.; Williams, P.J.; Reid, T.J.; Moerman, I.; Zhang, X.; van Daele, P.; Demeester, P. DFB laser with integrated wave-guided taper grown by shadow masked MOVPE. *Electron. Lett.* **1996**, *32*, 738–739. [CrossRef]

31. Kleijn, E.; Melati, D.; Melloni, A.; de Vries, T.; Smit, M.K.; Leijtens, X.J.M. Multimode interference couplers with reduced parasitic reflections. *Photonics Technol. Lett.* **2014**, *26*, 408–410. [CrossRef]

32. Bukkems, H.G.; Herben, C.G.P.; Smit, M.K.; Groen, F.H.; Moerman, I. Minimization of the loss of intersecting waveguides in InP-based photonic integrated circuits. *Photonics Technol. Lett.* **1999**, *11*, 1420–1422. [CrossRef]

33. D'Agostino, D.; Carnicella, G.; Ciminelli, C.; Ambrosius, H.P.M.M.; Smit, M.K. Low loss waveguides for standardized InP integration processes. In Proceedings of the IEEE Photonics Society Benelux, Enschede, The Netherlands, 2–4 November 2014.

34. Stabile, R.; Rohit, A.; Williams, K.A. Dynamic multi-path WDM routing in a monolithically integrated 8 × 8 cross-connect. *Opt. Exp.* **2014**, *22*, 435–442. [CrossRef] [PubMed]

35. Cheng, Q.; Stabile, R.; Rohit, A.; Wonfor, A.; Penty, R.V.; White, I.H.; Williams, K.A. First demonstration of automated control and assessment of a dynamically reconfigured monolithic 8 × 8 wavelength-and-space switch. *J. Opt. Commun. Netw.* **2015**, *7*, A388–A395. [CrossRef]

36. Stabile, R.; Williams, K.A. Optical power meter co-integrated with a fast optical switch for on-chip OSNR monitoring. In Proceedings of the 2015 International Conference on Photonics in Switching (PS), Florence, Italy, 22–25 September 2015.

37. Lee, J.H.; Jung, D.K.; Kim, C.H.; Chung, Y.C. OSNR monitoring technique using polarization-nulling method. *IEEE Photonics Technol. Lett.* **2001**, *13*, 88–90. [CrossRef]

38. Flood, E.; Guo, W.H.; Reid, D.; Lynch, M.; Bradley, A.L.; Barry, L.P.; Donegan, J.F. In-band OSNR monitoring using a pair of Michelson fiber interferometers. *Opt. Exp.* **2010**, *18*, 3618–3625. [CrossRef] [PubMed]

39. Morichetti, F.; Annoni, A.; Sorel, M.; Melloni, A. High-Sensitivity In-Band OSNR Monitoring System Integrated on a Silicon Photonics Chip. *IEEE Photonics Technol. Lett.* **2013**, *25*, 1939–1942. [CrossRef]

40. Li, Q.; Padmaraju, K.I.; Logan, D.F.; Ackert, J.J.; Knights, A.P.; Bergman, K. A Fully-integrated In-band OSNR Monitor using a Wavelength-tunable Silicon Microring Resonator and Photodiode. In Proceedings of the Optical Fiber Communications Conference and Exhibition (OFC), San Francisco, CA, USA, 9–13 March 2014.

41. Dasmahapatra, P.; Stabile, R.; Rohit, A.; Williams, K.A. Optical cross-point matrix using broadband resonant switches. *IEEE J. Sel. Top. Quantum Electron.* **2014**, *20*, 5900410. [CrossRef]

42. Varrazza, R.; Djordjevic, I.B.; Yu, S. Active vertical-coupler-based optical crosspoint switch matrix for optical packet-switching applications. *J. Lightwave Technol.* **2004**, *22*, 2034–2042. [CrossRef]

43. Guzzon, R.; Norberg, E.; Parker, J.; Johansson, L.; Coldren, L. Integrated InP-InGaAsP tunable coupled ring optical bandpass filters with zero insertion loss. *Opt. Exp.* **2011**, *19*, 7816–7826. [CrossRef] [PubMed]

44. Ji, R.; Yang, L.; Zhang, L.; Tian, Y.; Ding, J.; Chen, H.; Lu, Y.; Zhou, P.; Zhu, W. Five-port optical router for photonic networks-onchip. *Opt. Exp.* **2011**, *19*, 20258–20268. [CrossRef] [PubMed]

45. Stabile, R.; Dasmahapatra, P.; Williams, K.A. First 4 × 4 InP Switch Matrix Based on Third-Order Micro-Ring-Resonators. In Proceedings of the Optical Fibre Communication Conference (OFC), Anaheim, CA, USA, 20–24 March 2016.

46. Stabile, R.; DasMahapatra, P.; Williams, K.A. Fast and energy efficient Micro-ring-resonator-based 4 × 4 InP Switch matrix. In Proceedings of the European Conference on Integrated Optics (ECIO), Warsaw, Poland, 18–20 May 2016.

47. Stabile, R.; Dasmahapatra, P.; Williams, K.A. 4 × 4 InP switch matrix with electro-optically actuated higher order micro-ring resonators. *IEEE Photonics Technol. Lett.* **2016**, *28*, 2874–2877. [CrossRef]

48. Van der Tol, J.; Zhang, R.; Pello, J.; Bordas, F.; Roelkens, G.; Ambrosius, H.; Thijs, P.; Karouta, F.; Smit, M. Photonic integration in indium-phosphide membranes on silicon. *IET Optoelectron.* **2011**, *5*, 218–225. [CrossRef]

49. Stabile, R.; Kjellman, J.O.; Williams, K.A. Design of an optical via for large scale monolithic multilayer PICs. In Proceedings of the International Workshop on Optical Wave and Waveguide Theory and Numerical Modelling (OWTNM), Warsaw, Poland, 20–21 May 2016.

50. Kjellman, J.Ø.; Stabile, R.; Williams, K.A. Broadband giant group velocity dispersion in asymmetric InP dual layer, dual width waveguides. *J. Lightwave Technol.* **2017**, *35*, 3791–3800. [CrossRef]

![applied sciences logo]

applied sciences

MDPI

Article

Fast Reconfigurable SOA-Based Wavelength Conversion of Advanced Modulation Format Data [†]

Yi Lin [1],*, Aravind P. Anthur [1], Sean P. Ó Dúill [1], Fan Liu [2], Yonglin Yu [2] and Liam P. Barry [1]

[1]   School of Electronic Engineering, Dublin City University, Dublin 9, Ireland; aravind.anthur@dcu.ie (A.P.A.); sean.oduill@dcu.ie (S.P.ÓD.); liam.barry@dcu.ie (L.P.B.)
[2]   Wuhan National Laboratory for Optoelectronics, Huazhong University of Science and Technology, Wuhan 430073, China; fliu@hust.edu.cn (F.L.); yonglinyu@mail.hust.edu.cn (Y.Y.)
*   Correspondence: yi.lin6@mail.dcu.ie; Tel.: +353-1-700-5883
†   This paper is an extended version of paper published in the The Optical Fiber Communication Conference and Exhibition, OFC 2017 held in Los Angeles, 19–23 March 2017.

Received: 31 August 2017; Accepted: 3 October 2017; Published: 10 October 2017

Abstract: We theoretically analyze the phase noise transfer issue between the pump and the wavelength-converted idler for a nondegenerate four-wave mixing (FWM) scheme, as well as study the vector theory in nonlinear semiconductor optical amplifiers (SOAs), in order to design a polarization-insensitive wavelength conversion system employing dual co-polarized pumps. A tunable sampled grating distributed Bragg reflector (SG-DBR) pump laser has been utilized to enable fast wavelength conversion in the sub-microsecond timescale. By using the detailed characterization of the SGDBR laser, we discuss the phase noise performance of the SGDBR laser. Finally, we present a reconfigurable SOA-based all-optical wavelength converter using the fast switching SGDBR tunable laser as one of the pump sources and experimentally study the wavelength conversion of the single polarization quadrature phase shift keying (QPSK) and polarization multiplexed (Pol-Mux) QPSK signals at 12.5-Gbaud. A wide tuning range (>10 nm) and less than 50 ns and 160 ns reconfiguration time have been achieved for the wavelength conversion system for QPSK and PM-QPSK signals, respectively. The performance under the switching environment after the required reconfiguration time is the same as the static case when the wavelengths are fixed.

Keywords: all-optical networks; wavelength conversion devices; tunable lasers

---

1. Introduction

The massive growth in the demand for bandwidth for multimedia services and interactive networks is shaping a new era for today's communication networks. Future networks need to offer bandwidth-hungry applications like telemedicine, IP-TV, virtual reality gaming, video-on-demand, and high-speed internet access, combined with guaranteed Quality of Service [1]. In optical networks that employ wavelength division multiplexed (WDM), the use of optical switching technologies on a burst or packet level, combined with advanced modulation formats, would achieve greater spectral efficiency and utilize the existing bandwidth more efficiently. All-optical wavelength converters which typically comprise of tunable pump lasers, nonlinear media, and the tunable optical filter, are expected to be one of the key components in these broadband networks. They can be used to interface different networks and potentially increase the capacity of a communication system [2,3]. The wavelength converters can be also used at the network nodes to avoid contention and to dynamically allocate wavelengths to ensure optimum use of fiber bandwidth [4].

Recently, significant work has been undertaken on wavelength conversion based on four-wave mixing (FWM) by using different nonlinear devices such as nonlinear semiconductor optical amplifiers (SOAs) [5], highly nonlinear fiber (HNLF) [6–8], periodically poled Lithium Niobate (PPLN) [9], and nonlinear waveguides [10–12]. Nonlinear SOAs give the best conversion efficiency among these nonlinear devices because they are active devices [13]. To perform the wavelength conversion of data signals where the information is encoded onto the phase and amplitude of the optical carrier, e.g., quadrature phase shift keying (QPSK) and 16-ary state quadrature amplitude modulation (16-QAM), the wavelength conversion process must preserve the amplitude and phase information. Therefore, FWM is required as the wavelength conversion process because FWM preserves the amplitude and phase information of the input signal. In addition, FWM is transparent to the different modulation formats where the information can be encoded in the amplitude, phase, and polarization of the optical carrier [14], and is also transparent to the signal baud rate (the wavelength conversion of the on-off keying (OOK) signal at 100 Gbaud having been demonstrated in [15]). Recently, a lot of work has been done on the wavelength conversion of data signals employing different modulated formats including (Differential) QPSK, 8-Phase Shift Keying (PSK), 16-QAM, and 64-QAM based on FWM in SOAs using single pump and dual pump wavelength conversion systems [16–27].

In-addition to phase and amplitude, optical communication networks are employing polarization multiplexed (Pol-Mux) modulation formats to achieve twice the spectral efficiency by encoding data on two orthogonal polarizations of the light. When data is present in two orthogonal polarizations, it is very important to ensure that FWM-based wavelength converters are transparent to Pol-Mux modulation formats. Vector theory of FWM in nonlinear media is well understood and utilized for achieving polarization-insensitive wavelength conversion [28]. Vector theory was studied and verified in nonlinear SOAs and this was utilized to design a polarization-insensitive wavelength conversion scheme for Pol-Mux modulation formats utilizing dual co-polarized pumps in our earlier work [29]. Utilizing the understanding from these phase noise and polarization studies, we designed the most efficient and impairment-free wavelength converter based on FWM. We use this design to study the dynamic characteristics of the FWM-based wavelength converter and present the results, when the data is communicated in bursts/packets.

One important issue that needs to be considered when using FWM for all-optical wavelength conversion of data signals, where the information is encoded onto the phase of the optical carrier, is the phase noise transfer issue between the pump and the converted idler. In [30], a simple relationship between the linewidth of the signal, pump, and converted idler in the FWM process was found by theoretical analysis and verified experimentally. For the case of the single pump FWM scheme shown in Figure 1a, the linewidth of the converted idler is the linewidth of the signal plus four times pump linewidth, and for the case of the dual pump FWM scheme shown in Figure 1b, the linewidth of the converted idler is equal to the summed linewidths of the signal and the dual pumps. The phase noise transfer problem will cause a significant effect for the FWM-based wavelength conversion of data with phase encoding, as it becomes more difficult to recover the data with the induced uncertainty in the phase of the wavelength converted idler. Thus, the linewidth is an important parameter for the pump laser to be employed in the FWM-based wavelength conversion system.

An example of a typical all-optical packet-switched network which employs wavelength conversion is illustrated in Figure 1c. IP packets enter the network through the edge router where they are retransmitted on a new wavelength to avoid contention [31]. By employing the all-optical wavelength converter in this network, it can enable rapid routing of the same wavelength channels from any direction to any direction, and easily avoid the contention in which different packets with the same wavelengths are trying to leave the edge router. The wavelength converter consists of fast tuning tunable lasers as the pumps, an SOA, and a tunable optical filter. The wavelength of the wavelength-converted signal through FWM can be precisely altered by tuning the wavelength of the pump sources, and after the nonlinear wavelength conversion process in the SOA, the converted signal is then filtered out by using a tunable optical filter and sent to the next network node. The switching speed of this SOA-based wavelength converter mainly depends on the tunable pump lasers and the tunable optical filter [32], but in this work we focus on the effect of the tunable laser. The laser wavelength tuning speed can be as low as several ns for various tunable laser designs such as the sampled grating distributed Bragg reflector (SG-DBR) laser [33], the digital super-mode distributed Bragg grating (DS-DBR) laser [34], and the modulated grating Y-branch (MGY) laser [35]. These lasers can also achieve a wide tuning range (>45 nm), a high side mode suppression ratio (SMSR > 35 dB), and large output power (>10 dBm). All the components in the wavelength converter (lasers and SOA) can be integrated and these results motivate the construction of a compact, optically-integrated, and rapidly reconfigurable all-optical wavelength converter.

Most of the previous research on wavelength conversion is undertaken by using static, fixed-wavelength lasers. However, wavelength converters with rapid, dynamic wavelength re-configurability would be required to bypass the power-hungry electronic switches with high latency in the switching nodes of future transparent optical networks. The main focus of this paper is thus to develop a wavelength converter comprised of an SOA as the nonlinear element and a fast-switching SG-DBR tunable laser as one of the pump sources. In Section 2, we initially discuss the theory of the phase noise transfer issue in FWM and the dual co-polarized pumping scheme. In Section 3, using the detailed characterization of the SGDBR laser, we discuss the phase noise of the SGDBR. In Section 4, we provide a complete set of experimental results for single polarization QPSK [36] and experimentally demonstrate rapid wavelength conversion of a PM-QPSK signal with the switching time of tens of nanoseconds using a fast-switching, tunable laser as one of the pumps in a dual wavelength pumping scheme. After the reconfiguration time, the performance under a switching scenario is the same as the case when the wavelengths are held static. The experimental results indicate that the incoming signal can be precisely and quickly converted to the required wavelength channel on a nanosecond timescale.

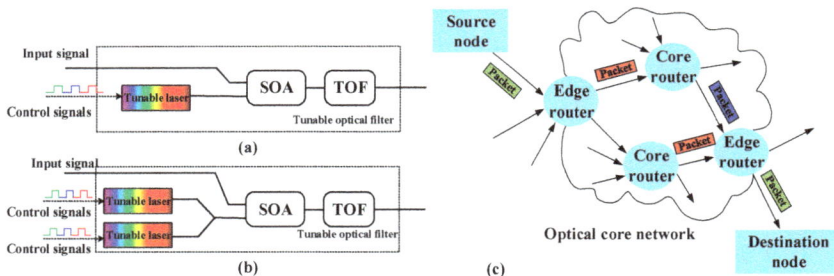

Figure 1. (**a**) A single pump semiconductor optical amplifier (SOA)-based wavelength converter employing one fast tunable pump laser. (**b**) A dual-pump SOA-based wavelength converter employing fast tunable pump lasers where the output wavelengths can be chosen by appropriately selecting the wavelength of the tunable lasers. The wavelengths of the tunable lasers are adjusted by the control signals to the SG-DBR devices. (**c**) Schematic of an all-optical packet-switched network.

2. Theory

2.1. Phase Noise and FWM

Consider two pumps with electric field intensities given by $\vec{E}_{p2}$ and $\vec{E}_{p1}$ mixing with a signal having an electrical field intensity of $\vec{E}_s$, generating an idler through FWM with a field intensity of E_i. Let the frequencies of the mixing pumps and the signal be represented as ω_{p1}, ω_{p2}, and ω_s, respectively. Let the phase of the mixing pumps and the signal be represented as ϕ_{p1}, ϕ_{p2}, and ϕ_s respectively. The frequency and the phase of the idler are related to the mixing frequencies by,

$$\omega_i = \omega_{p2} - \omega_{p1} + \omega_s \tag{1}$$

$$\phi_i = \phi_{p2} - \phi_{p1} + \phi_s \tag{2}$$

Unless both pumps have correlated phase noise, the phase noise of the idler is related to the phase noise of the mixing frequencies by the following relationship when the phase noise of the mixing frequencies is uncorrelated [16,30],

$$\Delta\sigma_i^2 = \Delta\sigma_{p1}^2 + \Delta\sigma_{p2}^2 + \Delta\sigma_s^2 \tag{3}$$

where $\Delta\sigma^2$ represents the variance of the phase noise. For white frequency phase noise, the linewidth (represented by $\Delta\omega$) and the phase error variance are linearly related [37]. Therefore, the linewidth of the idler due to white phase noise is related to the linewidth of the mixing pumps and the signal by,

$$\Delta\omega_i = \Delta\omega_{p1} + \Delta\omega_{p2} + \Delta\omega_s \tag{4}$$

Thus it can be observed from Equation (4) that the phase noise of the idler will be the sum of the phase noise of the mixing frequencies.

2.2. FWM with Dual Co-Polarized Pumps

We next look at the vector theory of FWM in nonlinear SOA. The states of polarization of the two co-polarized pumps and signal are shown in Figure 2a. Pump1 and pump2 are co-polarized. For simplicity, we consider a linearly polarized input signal at an arbitrary angle θ with respect to the pumps. The output spectrum of this FWM scheme is given in Figure 2b. The converted idlers generated by the beating between the dual co-polarized pumps have optical frequencies $\omega_{p1} - \omega_{p2} + \omega_s$ and $\omega_{p2} - \omega_{p1} + \omega_s$. We show that the FWM wavelength conversion scheme preserves the polarization properties of the signal within the idler at frequency $\omega_{p2} - \omega_{p1} + \omega_s$. The electric field of the two co-polarized pumps ($\vec{E}_{p1}$ and $\vec{E}_{p2}$) and signal ($\vec{E}_s$) are given by,

$$\vec{E}_{p1} = A_{p1} \exp\left(j(\omega_{p1}t + \phi_{p1})\right)\vec{x} \tag{5}$$

$$\vec{E}_{p2} = A_{p2} \exp\left(j(\omega_{p2}t + \phi_{p2})\right)\vec{x} \tag{6}$$

$$\vec{E}_s = A_s \exp\left(j(\omega_s t + \phi_s)\right)\left[\cos(\theta)\vec{x} + \sin(\theta)\vec{y}\right] \tag{7}$$

where A_{p1}, A_{p2}, and A_s are the amplitudes of pump1, pump2, and the signal, respectively. The idler at a frequency of $\omega_i = \omega_{p2} - \omega_{p1} + \omega_s$ is generated by the resultant of two beating processes given by [38–40],

$$\vec{E}_i = r(\omega_{p2} - \omega_{p1})\left(\vec{E}_{p2} \cdot \vec{E}_{p1}^*\right)\vec{E}_s + r(\omega_s - \omega_{p1})\left(\vec{E}_s \cdot \vec{E}_{p1}^*\right)\vec{E}_{p2}, \tag{8}$$

where $r(\omega_{p1} - \omega_{p2})$ is the efficiency of the beating process between the two pumps and $r(\omega_s - \omega_{p1})$ is the efficiency of the beating between the signal and the pump1 [41].

In the case of co-polarized pumps scheme, we tune the wavelengths of the two pumps such that the input signal is far enough from the pumps in order to avoid their interaction, which means $|\omega_{p2} - \omega_{p1}| \ll |\omega_s - \omega_{p1}|$. Therefore, the electrical field of the idler can be given by,

$$\vec{E}_i = r(\omega_{p2} - \omega_{p1})\left(\vec{E}_{p2} \cdot \vec{E}_{p1}^*\right)\vec{E}_s \tag{9}$$

Substituting Equation (5) to Equation (7) in Equation (9) gives,

$$\vec{E}_i = r(\omega_{p2} - \omega_{p1}) A_{p1}^* A_{p2} \vec{E}_s \exp\left[j(\omega_{p2} - \omega_{p1} + \omega_s)t + (\phi_{p2} - \phi_{p1} + \phi_s)\right] \tag{10}$$

It can be observed from Equation (10) that the generated idler has the same polarization as that of the input signal, and the output power is independent of the state of the polarization of the input signal. Assuming the SOA is polarization independent, the dual co-polarized pumping scheme can be used to convert a signal with arbitrary polarization with a constant conversion efficiency and single nonlinear SOA.

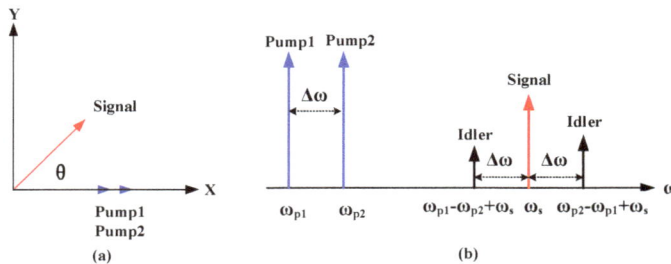

Figure 2. Four-wave mixing (FWM) scheme with dual co-polarized pumps. (**a**) The X and Y axes denote orthogonal states of polarization. In the dual pump FWM scheme, the polarization state of the wavelength-converted idler is that of the input signal provided both pumps have the same polarization. (**b**) Spectral locations of the pumps, signal, and idlers.

3. Linewidth Characterization of SGDBR Pump Lasers for Wavelength Conversion

Among the tunable lasers, the SGDBR laser has been proven to be a suitable fast tunable laser candidate [42], and the structure of this type of laser is shown in Figure 3a. It is comprised of one gain section, one phase section, and two grating sections. Due to the period in which the two gratings are sampled differently, each grating has different period of reflection maxima. Therefore, the Vernier effect is enabled and enhances the overall tuning range considerably. The function of the phase section is to allow the cavity modes to be shifted independently of the grating's reflection peak, achieving continuous tuning. As the data is modulated on the phase of the laser in the transmission systems employing advanced modulation formats, the phase noise of the lasers becomes an important factor in determining the transmission performance [43]. For the SGDBR laser, both the gain section and the passive tuning sections can contribute to the overall phase noise of the laser [44]. In order to characterize the phase noise of the SGDBR laser, the frequency modulation (FM) noise spectrum, which has been proved to be a very suitable measurement of the phase noise of lasers to be employed in coherent systems [45], can be obtained by the technique using a coherent receiver with a narrow-linewidth LO [37]. As shown in Figure 3b, white noise from the gain section and low frequency carrier noise from the passive tuning sections are observed. Due to the advance in digital signal processing (DSP), the low frequency excess phase noise can be compensated. A 2nd-order PLL scheme has been used in the carrier phase estimation to track the excess phase noise [46], which presents the ability to employ fast tunable laser for higher order modulation formats.

Apart from that, the phase noise of the SGDBR laser is strongly dependent on the injection currents on each section, and the linewidth varies between several 100's kHz to several MHz with different injected currents. The output wavelength and phase noise of the SGDBR laser as a function of the injected currents in the front section are measured and shown in Figure 4, with 90 mA current on the gain section and 0 mA current on the phase and back section. The white noise and carrier noise are represented by the high frequency linewidth and low frequency linewidth, which are calculated from the FM-noise spectrum by the simple equation: $\Delta \nu = \pi \cdots S(f)$ [47]. The high frequency linewidth is related to the FM noise level in the frequency range beyond 500 MHz, whereas the low frequency linewidth contributed by the passive tuning sections is extracted by the mean value of the FM-noise in the frequency range under 50 MHz. It can be observed from Figure 4, by changing the currents into a grating section, that the entire reflection comb of the grating section shifts in wavelength and the laser wavelength jumps not only to an adjacent longitudinal mode, but can also jump by several nm to another super-mode at a wavelength where the reflection peaks of the Vernier-tuned gratings have re-aligned. By increasing the currents on the grating section, the low frequency linewidth and high frequency linewidth are found to present an opposite trend. The low frequency linewidth usually increases until the wavelength jump occurs, while the trend for the high frequency linewidth is decreased. As most low frequency phase noise can be compensated by the DSP, the operation points of the SGDBR laser with less high frequency noise are preferred to perform data transmission experiments. According to these linewidth measurements results, two wavelengths of the SGDBR laser are chosen to switch between for the later wavelength conversion experiment. The 1548.68 nm wavelength of the SGDBR, which is represented by the black point A shown in Figure 4, is chosen by applying 90 mA current on the gain section while the other passive tuning sections are terminated. The 1553.70 nm wavelength (point B) is chosen by increasing the current on the front section to 3.4 mA.

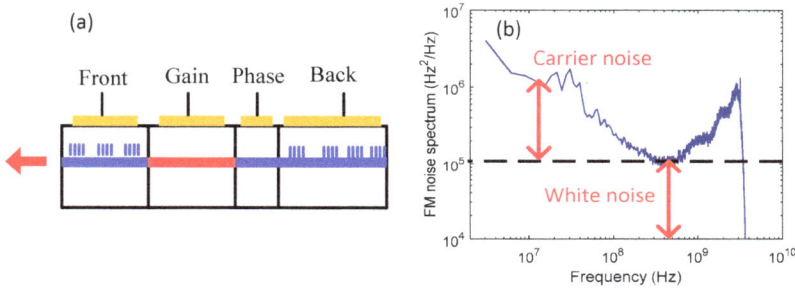

Figure 3. (**a**) Structure of the sampled, grating-distributed Bragg reflector (SGDBR) laser; (**b**) frequency modulation (FM) noise spectrum of the SGDBR laser.

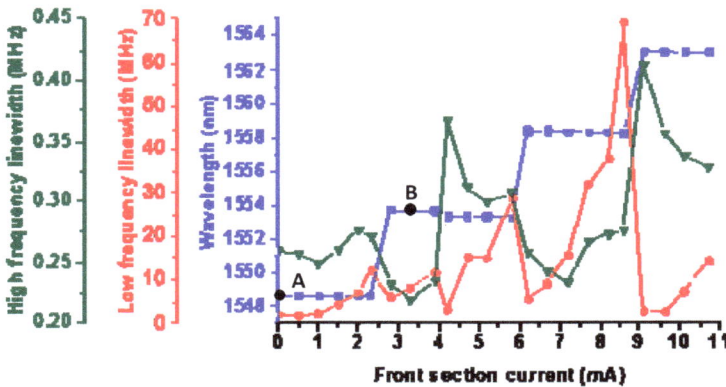

Figure 4. Measured high frequency linewidth (green triangles), low frequency linewidth (red dots), and output wavelength (blue squares) with different currents on the front section of the SGDBR laser, with 90 mA current on the gain section, 0 mA current on the phase and back section.

4. Experimental Setup for Reconfigurable Wavelength Converter

Figure 5 depicts the schematic diagram of the experimental setup of the SOA-based wavelength conversion of QPSK and PM-QPSK (blue dashed) signals using a SGDBR laser as one of the pump sources. A narrow linewidth external cavity laser (ECL) tuned to 1542.5 nm for all experiments is used as the signal source and was modulated with QPSK data at 12.5-Gbaud using an optical IQ modulator. The optical modulator is driven by electrical signals generated by the arbitrary waveform generator (AWG) with two uncorrelated pseudo-random bit sequences (PRBS) of 2^7-1 bits periodicity. The AWG operated at 25 GSa/s, which gives 2 samples per symbol for the 12.5-Gbaud QPSK signal. The two QPSK signals are amplified by using the radio frequency (RF) amplifiers. To implement the PM-QPSK signal scheme, the PM-QPSK signal is generated by using a Pol-Mux emulator, consisting of a polarization beam splitter (PBS) at its input, a passive stage with a delay of 4.88 ns (61 symbols), two polarization controllers, and a polarization beam combiner (PBC) to combine the polarization tributaries. The output of the two pumps are passed through polarization controllers, combined, and polarization-aligned using a 3 dB coupler and a PBS. For the wavelength conversion of the QPSK signal data, the generated optical QPSK signal is then coupled with two pumps (one SGDBR laser and one narrow linewidth ECL) and sent into the SOA-based wavelength converter. For both QPSK and PM-QPSK schemes, the power of the signal and the pumps at the input of the SOA are set at −10 dBm and 0 dBm, respectively, for optimum conversion efficiency [48]. The signal wavelength undergoes wavelength conversion through non-degenerate FWM in the SOA operating at 500 mA bias current. The SOA device used in the experiment is a nonlinear SOA with a very fast gain recovery time less than 25 ps. It operates over C-band with a typical small gain of 25 dB, a saturation power of +13 dBm, and less than 0.5 dB polarization-dependent saturated gain.

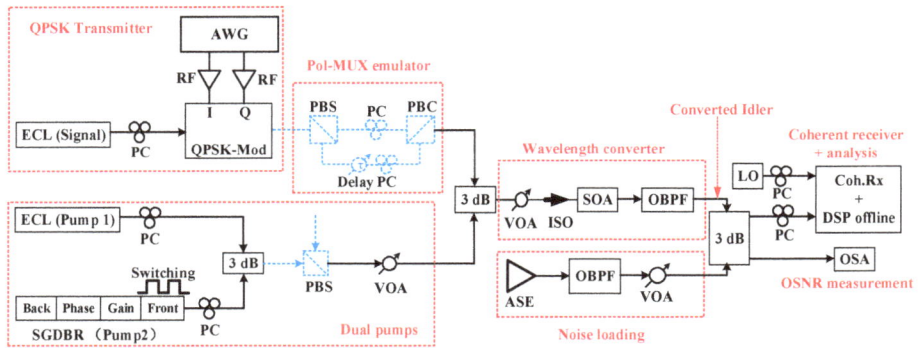

Figure 5. Schematic diagram of the reconfigurable SOA-based wavelength conversion of the quadrature phase shift keying (QPSK) and polarization multiplexed (Pol-Mux)-QPSK (blue dashed) signals employing a sampled grating distributed Bragg reflector (SGDBR) pump laser. PC: polarization controller, PBS: polarization beam splitter, PBC: polarization beam combiner, OBPF: optical band-pass filter, ISO: isolator, ASE: noise source, VOA: variable optical attenuator, OSA: optical spectrum analyzer, LO: local oscillator, Coh.Rx: coherent receiver.

After the FWM process in the SOA, the converted idler is then filtered out by using a tunable optical bandpass filter (OBPF). For system performance evaluation purposes, the optical signal to noise ratio (OSNR) of the idler is changed by adding amplified spontaneous emission (ASE) from an Erbium-doped fiber amplifier (EDFA) that is passed through a 2 nm bandwidth tunable optical bandpass filter. The filtered idler is then passed through a 3 dB splitter with one arm sent to the optical spectrum analyzer (OSA)for the OSNR measurement, and the other arm is passed into the polarization diversity coherent optical receiver and captured by a real-time oscilloscope sampling at 50 GSa/s for offline DSP processing. The received idler power at the input of the coherent receiver is maintained at −19 dBm. The data captured from the real-time oscilloscope is first resampled to 2 samples per symbol using a priori knowledge of the clock frequency. Then the constant modulus algorithm (CMA) [49–51] is utilized to enable polarization de-multiplexing for PM-QPSK signal. An *M*th power frequency offset compensation method [52–54] is employed to compensate the frequency offset between the converted idler signal and the local oscillator in the coherent receiver, with $m = 4$ being the number of distinct phases in the QPSK symbol set. In order to make this frequency offset compensation method work correctly, it is essential to make sure that the absolute frequency offset is less than Rs/(2 M), where Rs is the symbol rate. A second order Phase-Locked Loop (PLL) is employed for the phase noise estimation [46], and the synchronization is achieved by adding training symbols at the beginning of the data [55] in order to carry out the BER calculation.

5. Results and Discussion

The wavelength of the fast-tuning SGDBR pump laser is switched between two operating modes by applying a switching signal to the wavelength tuning sections. In order to benchmark the system performance, we first present the BER performances when pump2 is tuned and fixed to the two wavelengths that pump2 will later be dynamically switched between. The wavelength and power of the signal and pump1 (ECL) are kept constant throughout. The input and output spectra of the SOA when the SGDBR laser is set at 1548.68 nm and 1553.70 nm are shown in Figure 6a,b, respectively, and it can be observed that the indicated idler of interest has changed wavelength position from Ch1 (1541.395 nm) to Ch2 (1538.684 nm).

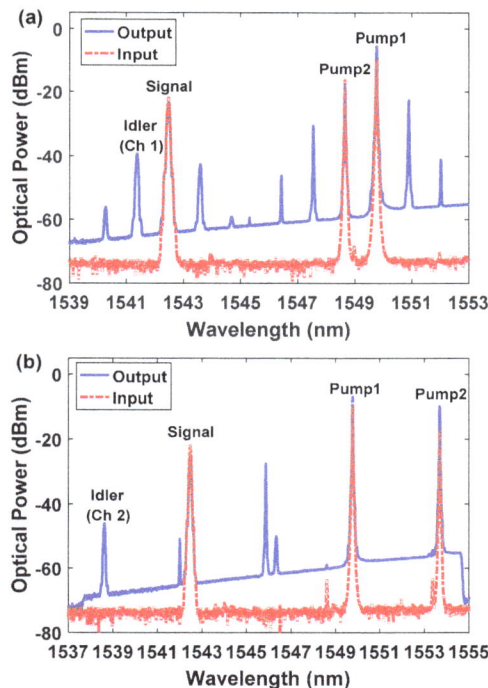

Figure 6. Input and output spectra of SOA showing the spectral locations of the signal of the external cavity laser (ECL), pump1 (ECL), pump2 (SGDBR), and converted idlers. (**a**) SOA input and output spectra when SGDBR is set at 1548.68 nm. (**b**) SOA input and output spectra when SGDBR is set at 1553.70 nm. The detected idlers are indicated.

BER performance as a function of OSNR at the receiver for the original signal; the signal after SOA and converted signals (idlers) for the QPSK and PM-QPSK signals at 12.5 GBaud are displayed in Figure 7a,b, respectively. It can be observed that the penalty between the original signal, the signal after SOA, and wavelength converted idler is under 0.5 dB for both cases, indicating the quality of the wavelength conversion scheme and potential usefulness. The constellation diagrams of the wavelength converted idler for a received OSNR of 12 dB and 14.5 dB are also given in Figure 7a,b, respectively. The phase noise of the FWM components is studied by using a coherent phase noise measurement technique [37,46]. The ECL's employed for the signal and pump1 have a linewidth of around 40 kHz and the linewidth of the 1548.68 nm and 1553.70 nm wavelengths of the SGDBR laser (employed as the second pump for FWM scheme) are 260 kHz and 230 kHz, respectively. The linewidth (from the high frequency phase noise region) of the idlers is measured to be around 370 kHz and 300 kHz, as expected when the SGDBR is set to the two operating modes, since the idler linewidth is the sum of the linewidths of the pumps and signal, and in this case the linewidth of the SGDBR pump laser dominates.

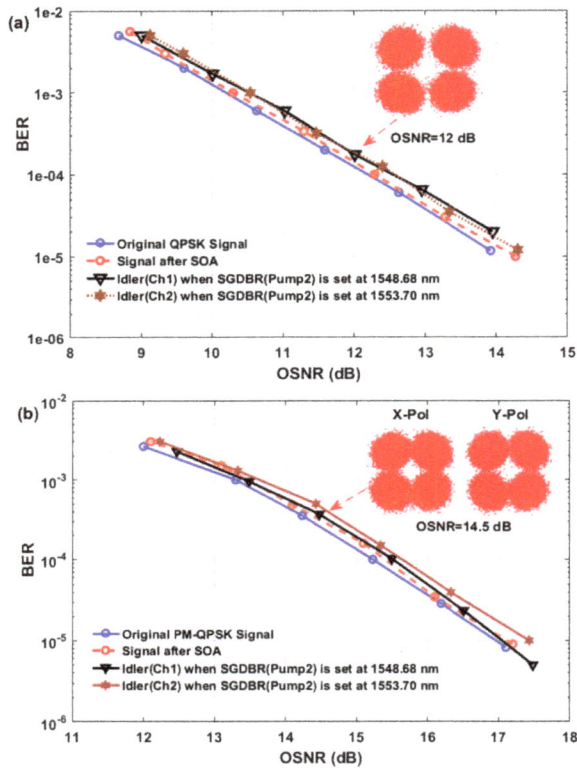

Figure 7. (**a**) Bit error rate (BER) versus optical signal to noise ratio (OSNR) curves for the input original signal, the signal after SOA, and the converted idlers for 12.5 GBaud QPSK signal. (**b**) BER versus OSNR curves for the input original signal, the signal after SOA, and the converted idlers for 12.5 GBaud PM-QPSK signal.

To calculate the limitations of the wavelength conversion scheme, we study the OSNR of the idler as a function of the detuning between the two pumps. The different wavelength converted idlers are filtered out by using an optical tunable band-pass filter. The OSNR measurement is then undertaken by using the OSA. In order to estimate the signal power and the noise floor correctly, the solution is to take two consecutive sweeps of the OSA with different resolution bandwidth (RBW) settings [56]. For the first sweep, OSA 0.2 nm RBW is used for the measurement of the signal power, and the second sweep measures the noise power by using 0.1 nm RBW setting. The output OSNR of the idler wavelength for the case with signal fixed at 1542.5 nm, pump1 fixed at 1549.8 nm, and the pump2 (SGDBR) tuned from 1548.7 nm to 1564.7 nm is displayed in Figure 8, which shows a tuning range of around 14 nm can be achieved with more than 9 dB OSNR, which is enough for the wavelength conversion of the QPSK data at 12.5 Gbaud to get a BER value below the 7% FEC limit (3.8×10^{-3}). Around 11 nm tuning range can be achieved for the wavelength conversion of PM-QPSK data to get a BER value below the 7% FEC limit. In order to investigate the time-resolved BER [57,58] performance of the fast-reconfigurable wavelength converter, we apply a square wave current with 500 kHz repetition rate to the front section of the SGDBR to switch the wavelength converted idler between Ch1 (1541.395 nm) and Ch2 (1538.684 nm), with the other currents applied to the SGDBR laser held constant and the received OSNR set at 12 dB for the QPSK signal and 14.5 dB for the PM-QPSK signal. The time-resolved BER measurement is characterized for Ch1 by fixing the center frequency of the OBPF to Ch1 and adjusting the LO to the wavelength of Ch1. The process is then repeated for Ch2.

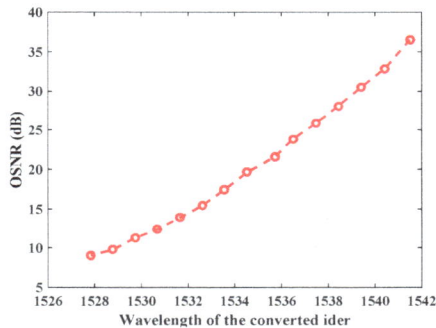

Figure 8. OSNR as a function of the wavelength of the converted idler.

We now present results of wavelength conversion when pump2 is dynamically tuned between 1547.68 nm and 1553.7 nm. In order to accurately estimate the time-resolved BER performance after a switching event, a number of switching events are captured via multiple acquisitions using the real-time scope. Each data point in the time-resolved BER curves corresponds to the probability of receiving an error in a 10 ns period. This means the BER is averaged over a block length of 125 symbols (10 ns), and data captured from 200 switching events is used for the calculation, with a total of 5×10^4 bits and 10^5 bits used for calculating each BER point in the time-resolved BER curves for QPSK and PM-QPSK signal at 12.5 GBaud, respectively. It can be seen from Figure 9a,b that the reconfiguration time (time to have a BER better than the 7% FEC limit) after a switch is approximately 50 ns and 160 ns for QPSK and PM-QPSK signal, respectively. The red curves in Figure 9a,b show the temporal frequency offset between the idler and the local oscillator in the coherent receiver after a switching event. It takes around 70 ns for the frequency of the wavelength converted idler for QPSK signal to fully stabilize after a switch, and 100 ns for PM-QPSK signal.

Figure 10 shows the BER measurement as a function of OSNR in a switching environment with different waiting time after the idler is switched to Ch1 and Ch2 when using QPSK and PM-QPSK decoding. The blue and red curves show the results for the data with a waiting time of 50 ns after switching when using QPSK decoding for Ch1 and Ch2, respectively. The brown and green curves are for a 160 ns waiting time when using PM-QPSK decoding for Ch1 and Ch2, respectively. The pink and black curves show the bad performance with a 50 ns waiting time using PM-QPSK decoding for Ch1 and Ch2, respectively. The required waiting time when using PM-QPSK decoding is longer than QPSK decoding mainly due to the longer convergence time associated with the CMA method used for de-multiplexing the dual-polarization packets.

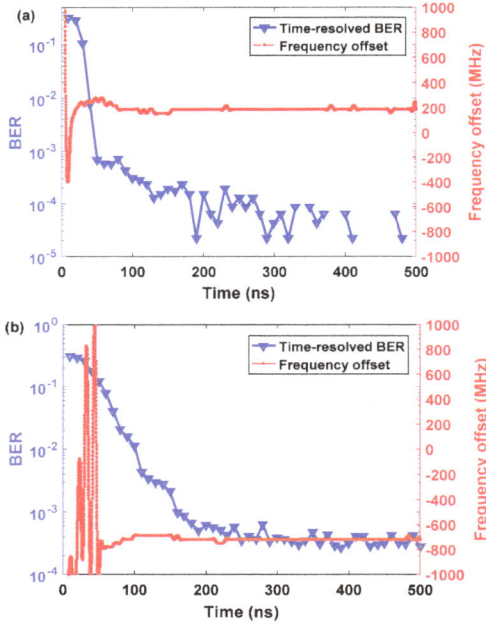

Figure 9. Time-resolved BER and frequency offset curves for: (**a**) QPSK signal; and (**b**) PM-QPSK signal at 12.5-GBaud, when the wavelength of the received signal (idler) is set at Ch1.

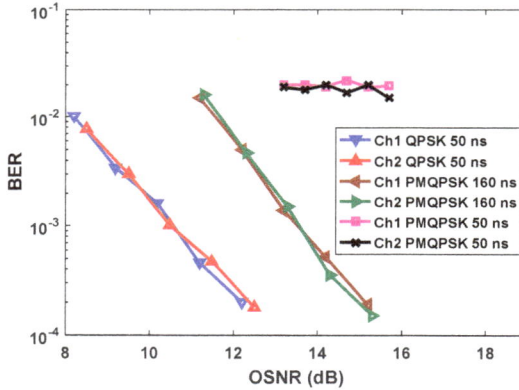

Figure 10. BER measurement as a function of OSNR in a switching environment with different waiting time after wavelength conversion when using QPSK and PM-QPSK decoding.

It also can be seen that the BER versus OSNR performance in a switching scenario corresponds with the static performance shown in Figure 7a,b, which indicates that the incoming signal can be precisely and quickly converted to the required channel on a timescale of around 50 ns and 160 ns for the 12.5-GBaud QPSK and PM-QPSK signal by using the wavelength converter we present.

6. Conclusions

The wavelength conversion of data with advanced modulation formats will play a significant role in next generation optical networks. FWM is one of the most advantageous technologies for the wavelength conversion of data employing advanced modulation formats. We theoretically study the phase noise transfer issue for a nondegenerate FWM scheme and the vector theory in nonlinear SOAs. According to these phase noise and polarization studies, we design the efficient and impairment-free SOA-based wavelength converter based on FWM. We demonstrate a rapidly reconfigurable SOA-based FWM wavelength conversion system using a fast-switching, tunable SGDBR laser as one of the pumps and experimentally study the wavelength conversion of QPSK and Pol-Mul QPSK signals at 12.5-Gbaud, with total data rates of 25 Gbps and 50 Gbps, respectively, using the proposed scheme. A wide tuning range (>10 nm) and a fast wavelength conversion time under 50 ns and 160 ns have been achieved for the proposed reconfigurable wavelength conversion system for QPSK and PM-QPSK signals, respectively. The reconfiguration time is mainly affected by the combination of the switching time of the tunable pump laser and the CMA convergence time in DSP. The performance under the switching environment after the required reconfiguration time is the same as the static case when the wavelengths are fixed, which makes it feasible to develop fast reconfigurable wavelength converters for dynamic, adaptive, and bandwidth-efficient optical networks by using rapid switching tunable pump lasers in conjunction with fast tuning optical filters.

Acknowledgments: This work has been supported by Science Foundation Ireland through grant numbers 13/RC/2077, 12/RC/2276, 15/US-C2C/I3132, and the HEA PRTLI 4 INSPIRE Programmes.

Author Contributions: L.P.B. conceived the initial ideas; Y.L., A.P.A., and F.L. designed and performed the experiments; Y.L., A.P.A., and S.P.Ó.D. wrote the paper. Y.Y. and L.P.B. reviewed the results and the manuscript.

Conflicts of Interest: The authors declare no conflict of interest.

References

1. Chang, G.K.; Yu, J.; Yeo, Y.K.; Chowdhury, A.; Jia, Z. Enabling technologies for next-generation optical packet-switching networks. *Proc. IEEE* **2006**, *94*, 892–910. [CrossRef]
2. Elmirghani, J.M.H.; Mouftah, H.T. All-optical wavelength conversion: Technologies and applications in DWDM networks. *IEEE Commun. Mag.* **2000**, *38*, 86–92. [CrossRef]
3. Bhopalwala, M.; Rastegarfar, H.; Kilper, D.C.; Wang, M.; Bergman, K. Energy efficiency of optical grooming of QAM optical transmission channels. *Opt. Express* **2016**, *24*, 2749–2764. [CrossRef] [PubMed]
4. Bayvel, P. *Wavelength Routing and Optical Burst Switching in the Design of Future Optical Network Architectures*; Proc. ECOC: Amsterdam, The Netherlands, 2001.
5. Govind, P. Agrawal, Population pulsations and nondegenerate four-wave mixing in semiconductor lasers and amplifiers. *J. Opt. Soc. Am. B* **1988**, *5*, 147–159.
6. Hu, H.; Jopson, R.M.; Gnauck, A.H.; Dinu, M.; Chandrasekhar, S.; Xie, C.; Randel, S. Parametric amplification wavelength conversion and phase conjugation of a 2.048-Tbit/s WDM PDM 16-QAM signal. *J. Lightwave Technol.* **2015**, *33*, 1286–1291. [CrossRef]
7. Li, C.; Luo, M.; He, Z.; Li, H.; Xu, J.; You, S.; Yang, Q.; Yu, S. Phase noise cancelled polarization-insensitive all-optical wavelength conversion of 557-Gb/s PDM-OFDM signal using coherent dual-pump. *IEEE J. Lightwave Technol.* **2015**, *33*, 2848–2854. [CrossRef]
8. Morshed, M.; Du, L.B.; Foo, B.; Pelusi, M.D.; Corcoran, B.; Lowery, A.J. Experimental demonstrations of dual polarization CO-OFDM using mid-span spectral inversion for nonlinearity compensation. *Opt. Express* **2014**, *22*, 10455–10466. [CrossRef] [PubMed]
9. Lu, G.; Albuquerque, A.; Puttnam, B.; Sakamoto, T.; Drummond, M.; Nogueira, R.; Kanno, A.; Shinada, S.; Wada, N.; Kawanishi, T. Pump-linewidth-tolerant optical wavelength conversion for high-order QAM signals using coherent pumps. *Opt. Express* **2014**, *22*, 5067–5075. [CrossRef] [PubMed]
10. Li, J.; O'Faolain, L.; Rey, I.H.; Krauss, T.F. Four-wave mixing in photonic crystal waveguides: Slow light enhancement and limitations. *Opt. Express* **2011**, *19*, 4458–4463. [CrossRef] [PubMed]

11. Ettabib, M.A.; Lacava, C.; Liu, Z.; Bogris, A.; Kapsalis, A.; Brun, M.; Labeye, P.; Nicoletti, S.; Syvridis, D.; Richardson, D.J.; Petropoulos, P. Wavelength conversion of complex modulation formats in a compact SiGe waveguide. *Opt. Express* **2017**, *25*, 3252–3258. [CrossRef] [PubMed]

12. Adams, R.; Spasojevic, M.; Chagnon, M.; Malekiha, M.; Li, J.; Plant, D.V.; Chen, L.R. Wavelength conversion of 28 GBaud 16-QAM signals based on four-wave mixing in a silicon nanowire. *Opt. Express* **2014**, *22*, 4083–4090. [CrossRef] [PubMed]

13. D'Ottavi, A.; Iannone, A.; Mecozzi, A.; Scotti, S.; Spano, P.; Dall'Ara, R.; Eckner, J.; Guekos, G. Efficiency and noise performance of wavelength converters based on FWM in semiconductor optical amplifiers. *IEEE Photonics Technol. Lett.* **1995**, *7*, 357–359.

14. Contestabile, G.; Banchi, L.; Presi, M.; Ciaramella, E. Investigation of transparency of FWM in SOA to advanced modulation formats involving intensity, phase, and polarization multiplexing. *J. Lightwave Technol.* **2009**, *27*, 4256–4261. [CrossRef]

15. Kelly, A.E.; Ellis, A.D.; Nesset, D.; Kashyap, R. 100 Gbit/s wavelength conversion using FWM in an MQW semiconductor optical amplifier. *Electron. Lett.* **1998**, *34*, 1955–1956. [CrossRef]

16. Anthur, A.; Watts, R.T.; O'Carroll, J.; Venkitesh, D.; Barry, L.P. Dual correlated pumping scheme for phase noise preservation in all-optical wavelength conversion. *Opt. Express* **2013**, *21*, 15568–15579. [CrossRef] [PubMed]

17. Naimi, S.T.; Duill, S.P.O.; Barry, L.P. Detailed investigation of the pump phase noise tolerance for the wavelength conversion of 16-QAM signals using FWM. *IEEE/OSA J. Opt. Commun. Netw.* **2014**, *6*, 793–800. [CrossRef]

18. Dúill, S.P.Ó.; Naimi, S.T.; Anthur, A.P.; Huynh, T.N.; Venkitesh, D.; Barry, L.P. Simulations of an OSNR limited wavelength conversion scheme. *IEEE Photonics Technol. Lett.* **2013**, *25*, 2311–2314. [CrossRef]

19. Naimi, S.T.; Dúill, S.P.Ó.; Barry, L.P. Simulations of the OSNR and laser linewidth limits for reliable wavelength conversion of DQPSK signals using four-wave mixing. *J. Opt. Commun.* **2014**, *310*, 150–155. [CrossRef]

20. Dúill, S.P.Ó.; Naimi, S.T.; Anthur, A.P.; Huynh, T.N.; Venkitesh, D.; Barry, L.P. Numerical generation of laser-resonance phase noise for optical communication simulators. *Appl. Opt.* **2015**, *54*, 3398–3406. [CrossRef] [PubMed]

21. Anthur, A.P.; Watts, R.T.; Zhou, R.; Anandarajah, P.; Venkitesh, D.; Barry, L.P. Penalty-free wavelength conversion with variable channel separation using gain-switched comb source. *Opt. Commun.* **2015**, *324*, 69–72. [CrossRef]

22. Anthur, A.P.; Watts, R.T.; O'Duill, S.P.; Zhou, R.; Venkitesh, D.; Barry, L.P. Impact of nonlinear phase noise on all-optical wavelength conversion of 10.7 GBaud QPSK data using dual correlated pumps. *IEEE J. Quantum Electron.* **2015**, *51*, 9100105. [CrossRef]

23. Filion, B.; Ng, W.C.; Nguyen, A.T.; Rusch, L.A.; LaRochelle, S. Wideband wavelength conversion of 16 Gbaud 16-QAM and 5 Gbaud 64-QAM signals in a semiconductor optical amplifier. *Opt. Express* **2013**, *21*, 19825–19833. [CrossRef] [PubMed]

24. Contestabile, G.; Yoshida, Y.; Maruta, A.; Kitayama, K. Coherent wavelength conversion in a quantum dot SOA. *IEEE Photonics Technol. Lett.* **2013**, *25*, 791–794. [CrossRef]

25. Krzczanowicz, L.; Connelly, M.J. 40 Gb/s NRZ-DQPSK data all-optical wavelength conversion using four wave mixing in a bulk SOA. *IEEE Photonics Technol. Lett.* **2013**, *25*, 2439–2441. [CrossRef]

26. Naimi, S.T.; Duill, S.P.Ó.; Barry, L.P. All Optical Wavelength Conversion of Nyquist-WDM Superchannels Using FWM in SOAs. *J. Lightwave Technol.* **2015**, *33*, 3959–3967. [CrossRef]

27. Dúill, S.P.Ó.; Barry, L.P. Improved reduced models for single-pass and reflective semiconductor optical amplifiers. *J. Opt. Commun.* **2015**, *334*, 170–173. [CrossRef]

28. Inoue, K. Polarization effect on four-wave mixing efficiency in a single-mode fiber. *IEEE J. Quantum Electron.* **1992**, *28*, 883–894. [CrossRef]

29. Anthur, A.P.; Zhou, R.; O'Duill, S.; Walsh, A.J.; Martin, E.; Venkitesh, D.; Barry, L.P. Polarization insensitive all-optical wavelength conversion of polarization multiplexed signals using co-polarized pumps. *Opt. Express* **2016**, *24*, 11749–11761. [CrossRef] [PubMed]

30. Hui, R.; Mecozzi, A. Phase noise of four-wave mixing in semiconductor lasers. *Appl. Phys. Lett.* **1992**, *60*, 2454–2456. [CrossRef]

31. Blumenthal, D.J.; Bowers, J.E.; Rau, L.; Hsu-Feng, C.; Rangarajan, S.; Wei, W.; Poulsen, K.N. Optical signal processing for optical packet switching networks. *IEEE Commun. Mag.* **2003**, *41*, S23–S29. [CrossRef]

32. Sadot, D.; Boimovich, E. Tunable optical filters for dense WDM networks. *IEEE Commun. Mag.* **1998**, *36*, 50–55. [CrossRef]

33. Jayaraman, V.; Mathur, A.; Coldren, L.A.; Dapkus, P.D. Theory design and performance of extended tuning range in sampled grating DBR lasers. *IEEE J. Quantum Electron.* **1993**, *29*, 1824–1834. [CrossRef]

34. Ward, A.J.; Robbins, D.J.; Busico, G.; Barton, E.; Ponnampalam, L.; Duck, J.P.; Whitbread, N.D.; Williams, P.J.; Reid, D.C.J.; Carter, A.C.; et al. Widely tunable DS-DBR laser with monolithically integrated SOA: Design and performance. *IEEE J. Sel. Top. Quantum Electron.* **2005**, *11*, 149–156. [CrossRef]

35. Wesström, J.-O.; Hammerfeldt, S.; Buus, J.; Siljan, R.; Laroy, R.; de Vries, H. Design of a widely tunable modulated grating Y-branch laser using the additive Vernier effect for improved super-mode selection. In Proceedings of the 18th International Semiconductor Laser Conference (ISLC), Garmisch, Germany, 29 September–3 October 2002.

36. Lin, Y.; Anthur, A.P.; O'uill, S.; Naimi, S.T.; Yu, Y.; Barry, L. *Fast Reconfigurable SOA-Based All-Optical Wavelength Conversion of QPSK Data Employing Switching Tunable Pump Lasers*; Proc. OFC: Los Angeles, CA, USA, 2017.

37. Kikuchi, K. Characterization of semiconductor-laser phase noise and estimation of bit-error rate performance with low-speed offline digital coherent receivers. *Opt. Express* **2012**, *20*, 5291–5302. [CrossRef] [PubMed]

38. Kyo, I. Polarization independent wavelength conversion using fiber four-wave mixing with two orthogonal pump lights of different frequencies. *J. Lightwave Technol.* **1994**, *12*, 1916–1920.

39. Jia, L.; Chen, L.; Dong, Z.; Cao, Z.; Wen, S. Polarization insensitive wavelength conversion based on orthogonal pump four-wave mixing for polarization multiplexing signal in high-nonlinear fiber. *J. Lightwave Technol.* **2009**, *27*, 5767–5774. [CrossRef]

40. Jonathan, L.; Mark, P.R.; Summerfield, A.; Madden, S.J. Tunability of polarization-insensitive wavelength converters based on four-wave mixing in semiconductor optical amplifiers. *J. Lightwave Technol.* **1998**, *16*, 2419–2427.

41. Zhou, J.; Park, N.; Vahala, K.J.; Newkirk, M.; Miller, B.I. Four-wave mixing wavelength conversion efficiency in semiconductor traveling-wave amplifiers measured to 65 nm of wavelength shift. *IEEE Photonics Technol. Lett.* **1994**, *6*, 984–987. [CrossRef]

42. Delorme, F. Widely tunable 1.55 μm lasers for wavelength-division-multiplexed optical fiber communications. *IEEE J. Quantum Electron.* **1998**, *34*, 1706–1716. [CrossRef]

43. Seimetz, M. *Laser Linewidth Limitations for Optical Systems with High-Order Modulation Employing Feed forward Digital Carrier Phase Estimation*; Proc. OFC: San Diego, CA, USA, 2008.

44. Jialin, Z.; Huijuan, Z.; Fan, L.; Yonglin, Y. Numerical Analysis of Phase Noise Characteristics of SGDBR Lasers. *J. Sel. Top. Quantum Electron.* **2015**, *21*, 1502009. [CrossRef]

45. Camatel, S.; Ferrero, V. Narrow linewidth CW laser phase noise characterization methods for coherent transmission system applications. *J. Lightwave Technol.* **2008**, *26*, 3048–3054. [CrossRef]

46. Huynh, T.N.; Nguyen, A.T.; Ng, W.C.; Nguyen, L.; Rusch, L.A.; Barry, L.P. BER Performance of Coherent Optical Communications Systems Employing Monolithic Tunable Lasers With Excess Phase Noise. *J. Lightwave Technol.* **2014**, *32*, 1973–1980. [CrossRef]

47. Di Domenico, G.; Schilt, S.; Thomann, P. Simple approach to the relation between laser frequency noise and laser line shape. *Appl. Opt.* **2010**, *49*, 4801–4807. [CrossRef] [PubMed]

48. Baveja, P.P.; Maywar, D.N.; Agrawal, G.P. Interband four-wave mixing in semiconductor optical amplifiers with ASE-enhanced gain recovery. *IEEE J. Sel. Top. Quantum Electron.* **2012**, *18*, 899–908. [CrossRef]

49. Liu, L.; Tao, Z.; Yan, W.; Oda, S.; Hoshida, T.; Rasmussen, J.C. *Initial Tap Setup of Constant Modulus Algorithm for Polarization De-Multiplexing in Optical Coherent Receivers*; Proc. OFC: San Diego, CA, USA, 2009.

50. Kikuchi, K. Performance analyses of polarization demultiplexing based on constant-modulus algorithm in digital coherent optical receivers. *Opt. Express* **2011**, *19*, 9868–9980. [CrossRef] [PubMed]

51. Leven, A.; Kaneda, N.; Chen, Y. *A Real-Time CMA-Based 10 Gb/s Polarization Demultiplexing Coherent Receiver Implemented in an FPGA*; Proc. OFC: San Diego, CA, USA, 2008.

52. Maher, R.; Millar, D.S.; Savory, S.J.; Thomsen, B.C. Widely Tunable Burst Mode Digital Coherent Receiver With Fast Reconfiguration Time for 112 Gb/s DP-QPSK WDM Networks. *J. Lightwave Technol.* **2012**, *30*, 3924–3930. [CrossRef]

53. Simsarian, J.E.; Gripp, J.; Gnauck, A.H.; Raybon, G.; Winzer, P.J. *FastTuning 224-Gb/s Intradyne Receiver for Optical Packet Networks*; Proc. OFC: San Diego, CA, USA, 2010.

54. Nakagawa, T.; Matsui, M.; Kobayashi, T.; Ishihara, K.; Kudo, R.; Mizoguchi, M.; Miyamoto, Y. *Non-Data-Aided Wide-Range Frequency Offset Estimator for QAM Optical Coherent Receivers*; Proc. OFC: Los Angeles, CA, USA, 2011.

55. Mori, Y.; Zhang, C.; Igarashi, K.; Katoh, K.; Kikuchi, K. Unrepeated 200-km transmission of 40-Gbit/s 16-QAM signals using digital coherent receiver. *Opt. Express* **2009**, *17*, 1435–1441. [CrossRef] [PubMed]

56. Dupre, J.; Stimple, J. Making OSNR Measurements in a Modulated DWDM Signal Environment. Available online: http://www.keysight.com/upload/cmc_upload/All/SLDPRE_2_OSNR_Measure.pdf (accessed on 8 October 2017).

57. O'Dowd, J.A.; Shi, K.; Walsh, A.J.; Bessler, V.M.; Smyth, F.; Huynh, T.N.; Barry, L.P.; Ellis, A.D. Time resolved bit error rate analysis of a fast switching tunable laser for use in optically switched networks. *J. Opt. Commun. Netw.* **2012**, *4*, A77–A81. [CrossRef]

58. Smyth, F.; Browning, C.; Shi, K.; Peters, F.; Corbett, B.; Roycroft, B.; Barry, L.P. *10.7Gbd DQPSK Packet Transmission Using a Widely Tunable Slotted Fabry-Perot Laser*; Proc. ECOC: Torino, Italy, 2010.

applied sciences

MDPI

Article

Application of Semiconductor Optical Amplifier (SOA) in Managing Chirp of Optical Code Division Multiple Access (OCDMA) Code Carriers in Temperature Affected Fibre Link

Md Shakil Ahmed * and Ivan Glesk

Department of Electronic and Electrical Engineering, University of Strathclyde, Glasgow G1 1XW, UK, ivan.glesk@strath.ac.uk
* Correspondence: shakil.ahmed@strath.ac.uk; Tel.: +44-745-901-1635

Received: 27 March 2018; Accepted: 30 April 2018; Published: 3 May 2018

Featured Application: The application of a semiconductor optical amplifier (SOA) in an incoherent Optical Code Division Multiple Access (OCDMA) system based on multi-wavelength picosecond code carriers is explored for mitigation of temporal distortion of an OCDMA auto-correlation affected by the temperature induced dispersion changes in a fibre optic transmission link.

Abstract: Chromatic and temperature induced dispersion can both severely affect incoherent high data rate communications in optical fibre. This is certainly also true for incoherent optical code division multiple access (OCDMA) systems with multi-wavelength picosecond code carriers. Here, even a relatively small deviation from a fully dispersion compensated transmission link can strongly impact the overall system performance, the number of simultaneous users, and the system cardinality due to the recovered OCDMA auto-correlation being strongly distorted, time-skewed, and having its full width at half maximum (FWHM) value changed. It is therefore imperative to have a simple tunable means for controlling fibre chromatic or temperature induced dispersion with high sub-picosecond accuracy. To help address this issue, we have investigated experimentally and by simulations the use of a semiconductor optical amplifier (SOA) for its ability to control the chirp of the passing optical signal (OCDMA codes) and to exploit the SOA ability for dispersion management of a fibre link in an incoherent OCDMA system. Our investigation is done using a 19.5 km long fibre transmission link that is exposed to different temperatures (20 °C and 50 °C) using an environmental chamber. By placing the SOA on a transmission site and using it to manipulate the code carrier's chirp via SOA bias adjustments, we have shown that this approach can successfully control the overall fibre link dispersion, and it can also mitigate the impact on the received OCDMA auto-correlation and its FWHM. The experimental data obtained are in a very good agreement with our simulation results.

Keywords: group velocity dispersion; chirp; optical code division multiple access auto-correlation; fibre propagation; chromatic dispersion; temperature induced dispersion; super-continuum generation; optical code division multiple access transmitter; optical code division multiple access receiver/decoder; semiconductor optical amplifier

1. Introduction

Optical fibre networks are a vital means of modern communication. Today's applications demand high data rate throughputs and this demand continues to increase. With incoherent data communication approaches, high data rates demand the use of short optical pulses as data carriers. An effective management of these short optical pulses, mainly their temporal broadening due to dispersion effects in optical fibre, is becoming increasingly challenging. It has been observed that the signals are 16 times more sensitive to chromatic dispersion at 40 Gb/s when compared to that at 10 Gb/s [1]. Therefore, to meet user demands for a much higher transmission bandwidth, the network operators face the challenge of how to adapt the existing networks to support the higher data rates that are demanded. However, without having simple ways to achieve tunable chromatic dispersion (CD) management, the current solution depends on accurate fibre length adjustments between different transmission fibre types and dispersion compensating fibre (DCF). The main disadvantage is the bulky nature of this approach. Another cost-effective and commonly used method of compensation is the use of fibre Bragg gratings (FBGs) [2]. Some other approaches include using new modulation schemes, chirp pre-compensation, electronic dispersion compensation, and the use of digital filters etc. [2–4].

Optical code division multiple access (OCDMA) is an advanced multiplexing scheme that is attractive for its high scalability and random access [5]. One of efficient and often used coding schemes by incoherent OCDMA systems is a two-dimensional wavelength-hopping time-spreading (2D-WH/TS) coding [6,7]. In this encoding approach, OCDMA sequences are spread simultaneously in both wavelength and time domains in order to improve the system scalability and its performance [5,6,8]. A 2D-WH/TS prime code is a class of two dimensional (2D: wavelength-time), incoherent asynchronous codes that support wavelength hopping within time-spreading over the Galois field of prime numbers with zero auto-correlation side-lobes and periodic cross-correlation functions of at most one [5]. Because 2D-WH/TS systems use multi-wavelength short picosecond code carriers, the data transmission is susceptible to the presence of chromatic dispersion in optical fibre [9]. Therefore, it is imperative to understand the influence of dispersion and to also have relatively simple means for dispersion mitigation. If untreated, the recovered OCDMA auto-correlation function (i.e., the received data pulse envelope) that is recovered by a CDMA receiver's decoder will be distorted and broadened due to code carriers' time-skewing [10], which is caused by chromatic dispersion in optical fibres. The multi-wavelength optical pulses that are used as OCDMA code carriers travel with different speeds since the fibre refractive index is wavelength dependent. Here also DCF modules are commonly used for CD compensation. However, the use of picosecond code carriers demands a high (sub-picosecond) dispersion compensation accuracy [11]. This is not a very practical approach if a short length of SMF-28 has to be added, requiring compensation by matching the length of the DCF fibres. There is therefore a widely recognized need for tunable dispersion compensation techniques.

Different types of tunable dispersion compensators, such as integrated tunable chirped fibre Bragg gratings (CFBG), adaptive tunable dispersion control, dispersion equalization by adjusting the equalizer to maximize the clock component, virtually imaged phase arrays (VIPA), micro electro mechanical systems (MEMS), and other methods were demonstrated [12–17]. The use of Semiconductor Optical Amplifiers (SOAs) has also been investigated for tunable dispersion control for use in incoherent OCDMA systems. In [18,19], an SOA that was located at the receiver site was investigated for its effectiveness in reshaping the OCDMA auto-correlation functions distorted by the dispersion. In [19], it was experimentally demonstrated that the influence of the fibre link temperature on the recovered OCDMA auto-correlation could be mitigated by introducing an SOA as part of the OCDMA receiver. In [20], an SOA that was located at the transmission end was used to control the chirp of OCDMA code carriers to mitigate the effect of CD on the OCDMA auto-correlation width changes at room temperature.

In this paper, to the best of our knowledge, we investigate for the first time the use of an SOA on a transmitter site to mitigate the effects of fibre temperature fluctuations that affect the received OCDMA auto-correlation that is recovered on the receiver site. We have experimentally shown, and by simulations, that a distorted OCDMA auto-correlation due to the temperature induced fibre dispersion (TD) can be corrected by manipulating the chirp of code carriers when traversing a biased SOA prior to entering the transmission link. In addition, the simultaneous effects of the varying values of group velocity dispersion (parameter β_2) and fibre link temperature on the OCDMA auto-correlation width changes were also investigated. We have shown that the impact of a changing β_2 (0.03–0.06) ps^2/nm, when the fibre link is exposed to 50 °C, can be continuously compensated by using the SOA controlling the code carriers' chirp. Our simulation results are in very good agreement with the experimental observations.

2. Experimental Setup

The experimental setup that we used for our investigation is shown in Figure 1. Here, the ps ML Laser produces a 1.8 ps single wavelength (1545 nm) optical clock at OC-48 rate (bit-width equal to ~400 ps), which is then passed through an optical supercontinuum generator (OSG) by PriTel, Inc. (Naperville, IL, USA). The OSG consists of a high-power erbium doped fiber amplifier and dispersion decreasing fiber (DDF). The optical clock from the ps ML Laser that was centered at 1545 nm after the amplification by an 18 dBm erbium doped fibre amplifier (EDFA) is injected into approximately a 1 km long DDF. This way, a 3.2 nm wide optical supercontinuum is generated in a spectral region of 1550–1553.2 nm. The supercontinuum is then spectrally sliced by an OCDMA encoder (OKI Industries, Saitama, Japan), which is based on four FBGs (central frequencies are: λ_1 = 1551.72 nm, λ_2 = 1550.92 nm, λ_3 = 1552.52 nm, λ_4 = 1550.12 nm) matching the 100 GHz ITU grid. The physical separation of individual FBGs is such that the encoder generates a following 2D-WH/TS OCDMA code: (1-λ_2, 21-λ_4, 24-λ_1, 39-λ_3). Here, the integer number leading λ_i represents the order of the time slot (called a chip) containing the carrier λ_i within the 400 ps long bit-width. This OCDMA code is then amplified by 15 dBm EDFA-1 and is transmitted through a 19.5 km long fibre spool that is located inside of the environmental chamber (SM-32C from Thermatron Industries, Holland, MI, USA).

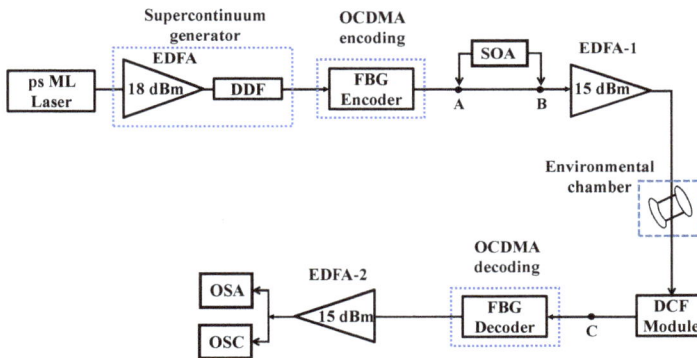

Figure 1. Experimental setup for the chirp control investigation by SOA at a transmitter site. ps ML Laser: picosecond mode locked laser; EDFA: erbium doped fibre amplifier; DDF: dispersion decreasing fibre; FBG: fibre Bragg gratings; DCF: dispersion compensating fibre; SOA: semiconductor optical amplifier; OSA: optical spectrum analyser; OSC: oscilloscope.

At the receiving end, an OCDMA auto-correlation is recovered from the received signal by an OCDMA decoder (OKI Industries) that is matched to the OCDMA encoder. In this process, the decoder removes all delays that are imposed on individual code carriers by the encoder [21]. The result is an OCDMA auto-correlation peak, i.e., which represents the received data. This is illustrated in Figure 2. To compensate for the code carriers' optical attenuation resulting from the fibre link attenuation, and also for the losses during the OCDMA auto-correlation recovery by the decoder, the resulted signal is amplified by a 15 dBm EDFA-2. For longer transmission distances, when the transmitted signal suffers more attenuation losses, an EDFA may be required before the OCDMA decoder. In our investigation, the decoded signal was then analysed using an optical spectrum analyser (OSA—Agilent 86146B, Agilent Technologies, Santa Clara, CA, USA) and oscilloscope, (OSC—Agilent Infiniium DCA-J 86100Cwith a 64 GHz optical sampling head, Agilent Technologies, Santa Clara, CA, USA). For future applications, to avoid the need for high bandwidth electronics, we assume using a hard-limiter in the front of the 2D-WH/TS OCDMA decoder that is similar to one demonstrated in [8]. This hard-limiter would be based on a picosecond all-optical time gating, followed by low-bandwidth electronics.

To study the effect of temperature on the transmitted OCDMA codes, the 19.5 km link was CD compensated at room temperature $T = 20\,°C$ using a commercially available dispersion compensating module DCF. In our investigations, the fibre link temperature will be varied from 20 °C to 50 °C. As a result, the code carriers will be affected by these temperature changes due to fibre refraction index temperature dependency [19]. The temperature induced dispersion coefficient D_T of optical fibre is negative. Based on [22], $D_T = -0.0016\,ps/nm\cdot km/°C$, therefore pulses of longer wavelengths travel with higher speed ν than those of shorter wavelengths (in our case $\nu_{\lambda4} < \nu_{\lambda2} < \nu_{\lambda1} < \nu_{\lambda3}$). This is contrary to CD where the chromatic dispersion coefficient $D_{CD} > 0$. The impact of D_T is twofold:

(1) It causes the rise of the so-called time-skewing effect illustrated in Figure 2c. We can see that because the code carriers having a longer wavelength travel faster than those with a shorter wavelength, instead of being placed on top of each other at the fibre receiving end after the OCDMA decoder (Figure 1), as is shown in Figure 2b, the code carriers are instead time shifted (Figure 2c). The resulting OCDMA auto-correlation envelope has thus become skewed.
(2) Because of finite code carriers' linewidth and D_T being negative, they will experience a temporal narrowing (squeezing) [19] (see Figure 2c). This result is opposite to the CD effect where the chromatic dispersion coefficient $D_{CD} > 0$, and it would therefore cause temporal broadening of the code carriers.

Our next step is to formally describe the above.

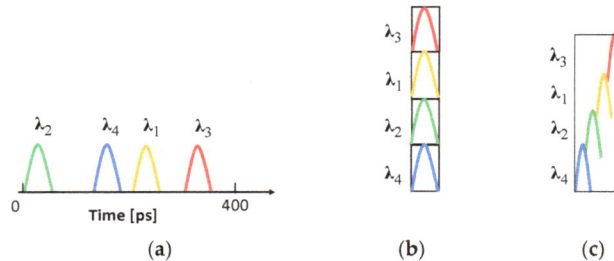

Figure 2. Illustration of the effect of temperature induced dispersion in fibre transmission link on incoherent optical code division multiple access (OCDMA). (**a**) Two-dimensional wavelength-hopping time-spreading (2D-WH/TS) transmitted OCDMA code; (**b**) Recovered OCDMA auto-correlation—ideal case; and, (**c**) Recovered OCDMA auto-correlation time-skewed and compressed due to the temperature induced dispersion effects ($D_T < 0$).

3. Analyses of Chirp and Temperature Induced Fibre Dispersion on Optical Code Division Multiple Access Auto-Correlation

We will now develop an equation describing the OCDMA auto-correlation envelope that was recovered by the OCDMA decoder on the receiver site being affected by the initial code carriers' chirp and fibre temperature changes.

A temporal Gaussian shape of an optical pulse affected by chirp C can be expressed as [23]:

$$P_L(t) = P_0 \exp\left[-2.77(1+iC)\left(\frac{t}{\tau}\right)^2\right], \tag{1}$$

where t is time, P_0 is a peak power and the coefficient 2.77 indicates that its width τ is measured at full width at half maximum (FWHM).

In our investigation we will use a 2D-WH/TS OCDMA system based on four multi-wavelength picosecond pulses as code carriers, each pulse carrier j has a line-width of $\Delta\lambda j$. Carriers are spectrally separated by $\Delta\Lambda j$ where $j = 1, 2, 3, 4$, respectively. Using Equation (1) and following [24] for the OCDMA auto-correlation temporal envelope that was recovered by the OCDMA receiver R_x, we can write the following expression:

$$S_{L(\lambda 1 - \lambda 4)}(t) = \sum_{k=0}^{w-1} P \exp\left[-2.77(1 + iC)\left(\frac{t - k\Delta\tau_0}{\tau - \Delta\tau_0} \right)^2 \right], \tag{2}$$

In the above expression, $w = 4$ is the number of code carriers (wavelengths) and it is also known as the code weight. The numerator $(t - k\Delta\tau_0)$ represents time skewing among wavelength code carriers and the denominator $(\tau - \Delta\tau_0)$ reflects the carrier width change due to temperature induced dispersion effects in the fibre link.

By following [20], we now develop a mathematical model for describing the effect of the code carriers' chirp and temperature induced dispersion changes in the fibre transmission link on the OCDMA auto-correlation envelope. Because we have used the same OCDMA hardware, we used in [20], the chirp value $C = 2.36$ at the OCDMA encoder output (point A, Figure 1) will be considered to be the same for the purpose of this investigation.

As has already been mentioned, the 19.5 km fibre transmission link was CD compensated at a room temperature of $T = 20\ °C$. In order to CD compensate the 19.5 km SMF-28 fibre, we applied a commercially available DCF module. Because of small dispersion mismatch between the link and DCF, this resulted in overall link overcompensation that could be represented by an average CD coefficient $D_{CD} = -0.047$ ps/nm·km. We found the fibre link overall group velocity dispersion (GVD) coefficient to be $\beta_2 = +0.06$ ps^2/km by following the expression as [23]:

$$\beta_2 = -(2\pi c/\lambda^2)/D_{CD}, \tag{3}$$

Now, by formally combining CD and TD effects, the expression for a code carrier pulse width change can be written as:

$$\Delta\tau_j = D_{CD} \times L \times \Delta\lambda_j + D_T \times \Delta\lambda_j \times L \times \Delta T, \tag{4}$$

where $D_{CD} = -0.047$ ps/(nm·km) represents the residual dispersion of the CD compensated fiber transmission link, $L = 19.5$ km is the link's length, $D_T = -0.0016$ ps/nm·km/°C [22], and ΔT is a fiber link temperature change. Because of using a 100 GHz FBG 2D-WH/TS OCDMA encoder for an optical supercontinuum slicing, values of $\Delta\lambda_j = \Delta\Lambda_j = 0.8$ nm.

At the OCDMA decoder, the frequency domain representation of the recovered OCDMA auto-correlation envelope, after $L = 19.5$ km of code propagation in the fibre link, can be obtained from Equation (2), by following [25] as:

$$S_{L(\lambda 1 - \lambda 4)}(f) = \text{fft}\{S_{L(\lambda 1 - \lambda 4)}(t)\} \times \{\exp(-i\,\omega_j^2 \times \beta_2/2) \times L\}, \tag{5}$$

again, $\beta_2 = +0.06$ ps^2/km is the fibre link GVD, ω_j is the angular frequency of the particular code carrier j, and $\{\exp(-i\,\omega_j^2 \times \beta_2/2) \times L\}$ factors in GVD effects that are imposed by the propagation in the fibre link at a given temperature T (note, the coefficient β_2 is temperature dependent).

Now, in order to obtain the OCDMA auto-correlation envelope S_L at the receiving end in the time domain, we need to perform an inverse Fourier transform (ifft) on Equation (5) after substituting Equations (2) and (5), we can formally write:

$$S_{L(\lambda_1 - \lambda_4)}(t) = \text{ifft}\{S_{L(\lambda_1 - \lambda_4)}(f)\}, \tag{6}$$

where for reasons indicated earlier we use $C = +2.36$. Based on the analyses carried out in [26], we conclude that larger C ($C > 0$) the code carriers have at the OCDMA transmitter output, the more broadening they will experience during fibre propagation. If the value of C can be appropriately reduced, the OCDMA auto-correlation broadening at the receiver site can be reduced or even eliminated.

The possibility of using the SOA for the code carriers chirp control to eliminate fibre temperature induced dispersion $D_T < 0$ will be investigated next, first theoretically using the developed Equation (6), and then experimentally by inserting an SOA between points A and B (see Figure 1). The temperature range in our investigation will be from 20 °C to 50 °C.

4. Comparison of Simulation with Experimental Results

The time-domain representation of the OCDMA USER-1 code is shown in Figure 3a. The experimentally obtained OCDMA auto-correlation that was recovered by the OCDMA decoder after OCDMA code travelled $L = 19.5$ km at room temperature $T = 20$ °C is shown in Figure 3b. The measured FWHM width was found to be 12 ps. This experiment was then repeated for $T = 50$ °C and the recovered OCDMA auto-correlation is shown in Figure 3c. We can see that the FWHM width has been broadened to 13 ps by this 30 °C increase in temperature. The corresponding results that are based on our simulations are shown in Figure 4. Figure 4a shows the simulated OCDMA auto-correlation envelope $S_{L=19.5;T=20}$ that was obtained when the fibre temperature was $T = 20$ °C and the code propagation distance $L = 19.5$ km. The calculated FWHM value is 12 ps, which is identical to the value that was obtained in our experiment (see Figure 3b). Similarly, simulations of the recovered OCDMA auto-correlation envelope $S_{L=19.5;T=50}$ for the temperature $T = 50$ °C is shown in Figure 4b. Here, the calculated FWHM value is 12.8 ps and it is in very good agreement with the experimentally observed 13 ps seen in Figure 3c.

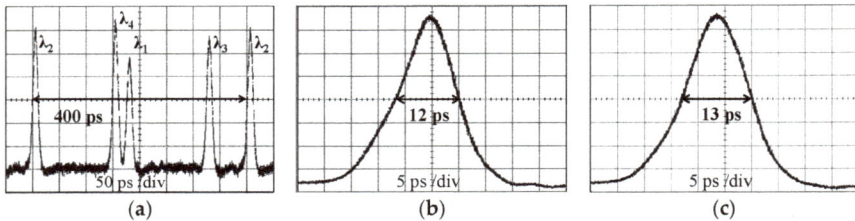

Figure 3. (**a**) OCDMA encoder output showing a user-1 (1-λ_2, 21-λ_4, 24-λ_1, 39-λ_3) 2D-WH/TS OCDMA code 400 ps long as seen by a bandwidth limited oscilloscope with 64 GHz optical sampling head; (**b**) Recovered OCDMA autocorrelation by the OCDMA decoder at the room temperature $T = 20$ °C (FWHM width = 12 ps); and, (**c**) at $T = 50$ °C (FWHM width = 13 ps), respectively.

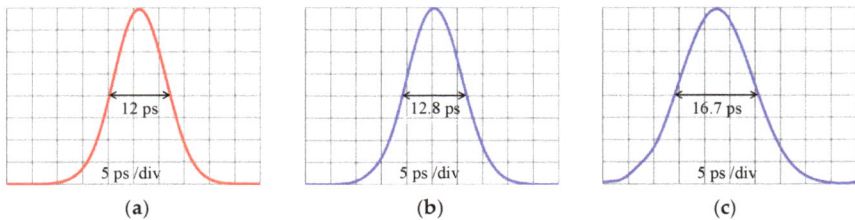

Figure 4. Simulations of the recovered OCDMA auto-correlation $S_{L;T}$ envelope by the OCDMA decoder: (**a**) for $L = 19.5$ km and $T = 20$ °C the FWHM width = 12 ps; (**b**) for $L = 19.5$ km and $T = 50$ °C the FWHM width = 12.8 ps; and, (**c**) for $L = 40$ km and $T = 80$ °C the FWHM width = 16.7 ps.

To further emphasise the severity of detrimental effects of temperature induced fiber dispersion, Figure 4c shows the recovered OCDMA auto-correlation envelope $S_{L=40;T=80}$ simulated for a fiber length $L = 40$ km when exposed to the temperature $T = 80$ °C. The corresponding FWHM value is 16.7 ps.

Our next step was to experimentally investigate a possibility of using the SOA to restore the OCDMA auto-correlation width that is affected by fibre temperature increases to its original value that was observed at room temperature, $T = 20$ °C. To conduct this investigation, we inserted an SOA (OPA-20-N-C, Kamelian, Glasgow, UK) at the transmission site between point A and B (see Figure 1). The role of the SOA will be to control the incoming chirp of the code carriers passing through the SOA by changing the bias current, I. This technique is known as a pre-chirping.

Figure 5 shows the experimentally obtained OCDMA auto-correlations that were recovered by the OCDMA decoder when the OCDMA code first passed the biased SOA and then travelled $L = 19.5$ km in the transmission link that was exposed at temperature $T = 50$ °C. The OCDMA auto-correlations that are shown in Figure 5a–c were obtained for SOA bias currents $I = 42$ mA, 14 mA, and 8 mA, respectively, and resulted in respective FWHM values of 14 ps, 13 ps, and 12 ps. The loss/gain of SOA w.r.t. the various bias currents have been shown in Figure 6.

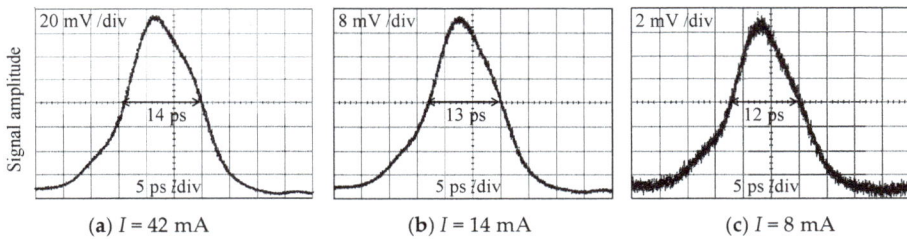

Figure 5. Experimentally recovered OCDMA auto-correlation by the OCDMA decoder for the fibre link temperature $T = 50$ °C and SOA bias current set to: (**a**) $I = 42$ mA (FWHM width = 14 ps); (**b**) $I = 14$ mA (FWHM width = 13 ps); and, (**c**) $I = 8$ mA (FWHM width = 12 ps), respectively.

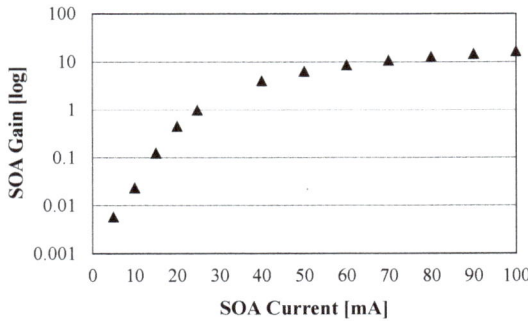

Figure 6. SOA Gain vs. Drive Current.

Because transmitting low signal levels can affect the system's performance and bit error rate (BER), an EDFA-1 on the transmitter site (see Figure 1) was used to amplify the signal.

Figure 7 shows the corresponding simulations of the experimental results that were shown before in Figure 5 i.e., the recovered OCDMA auto-correlation envelope $S_{L=19.5;T=50}$ for the OCDMA code propagation distance $L = 19.5$ km, fibre link temperature $T = 50$ °C, and chirp value: (a) $C = 5$ resulting in the FWHM width = 14.2 ps; (b) $C = 3$ resulting in the FWHM width = 13 ps; and, (c) $C = 1$ resulting in the FWHM width = 12 ps, respectively.

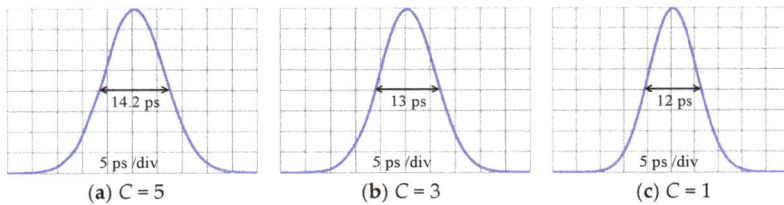

Figure 7. Simulations of the recovered OCDMA auto-correlation envelope $S_{L=19.5;T=50}$ for OCDMA code propagation distance $L = 19.5$ km, $T = 50\,°C$, and C equals to (**a**) 5 resulting FWHM width = 14.2 ps; (**b**) 3 resulting FWHM width = 13 ps; and, (**c**) 1 resulting FWHM width = 12 ps, respectively.

Next, we have investigated the cumulative effect of the varying value of the transmission link GVD (parameter β_2) on OCDMA auto-correlation FWHM's relative changes for the code transmission in an $L = 19.5$ km fibre link that was exposed to a temperature of $T = 50\,°C$. We have shown (see Figure 8) that the impact of changing β_2 (0.03–0.06) ps^2/nm (affected by $T = 50\,°C$) can also be compensated using the SOA by controlling the chirp of OCDMA code carriers.

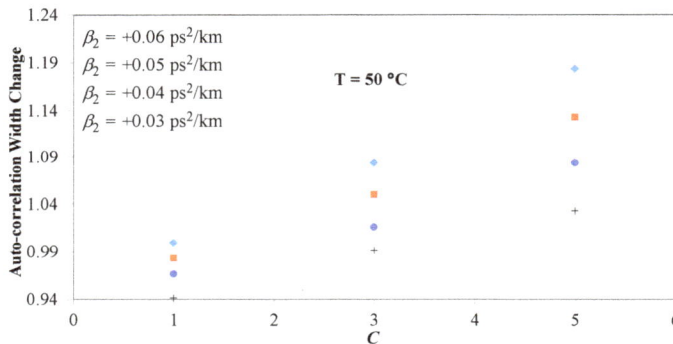

Figure 8. Compensation of the impact of β_2 on OCDMA auto-correlation width changes by using an SOA to control code carriers' chirp at $T = 50\,°C$.

One of the key performance indicators for incoherent OCDMA systems is a power ratio between the recovered auto-correlation peak and cross-correlation noise. In [27], it was shown that this ratio 5:1 would worsen to 2.6:1 due to 20 °C temperature fluctuations in fibre, thus strongly affecting the system BER [28]. As shown in [29], the OCDMA system transmission quality and the BER are affected by the pulse shape. As such, a skewed pulse profile of the recovered OCDMA auto-correlation will also impact the system BER and the overall system performance.

5. Discussion

Our recent work has been focused on the application of SOAs for improving the performance of incoherent OCDMA systems that are based on a 2D-WH/TS coding scheme using multi-wavelength picosecond code carriers. In [19], we reported for the first time the use of an SOA on the receiver site to continuously compensate for the broadening of the OCDMA auto-correlation due to the temperature induced dispersion in the optical fibre (the temperature induced dispersion coefficient $D_T < 0$ in SMF-28). The SOA compensation was applied directly on the distorted OCDMA auto-correlation composed of multi-wavelength code carriers sitting on top of each other. We have shown that the broadening of the OCDMA auto-correlation width due to the fibre temperature induced dispersion can be mitigated using the twin SOA by controlling its chirp through bias current adjustments once the OCDMA auto-correlation is recovered by the OCDMA decoder. In [18], we investigated the use of an SOA that was also placed on the receiver site for a continuous manipulation of the broadened OCDMA auto-correlation due to fibre link CD (chromatic dispersion coefficient $D_{CD} > 0$ in SMF-28). The SOA compensation was applied directly on the distorted OCDMA auto-correlation composed of four multi-wavelength code carriers sitting on top of each other. In this configuration, we had to use a picosecond optical pulse called a holding beam (OP/HB) synchronized with the OCDMA auto-correlation at the SOA input to be able to demonstrate that the OCDMA auto-correlation that was broadened by the fibre chromatic dispersion can be continuously 'compressed'. In [20], we reported the use of an SOA on the transmitter site (right after the OCDMA encoder) to compensate the OCDMA auto-correlation broadening due to the fibre chromatic dispersion ($D_{CD} > 0$), we observed on the OCDMA receiver site by controlling the chirp of the individual multi-wavelength OCDMA code carriers spread over a bit period before the code is launched into a fibre link.

In this paper, we have investigated for the first time how controlling the chirp of the OCDMA code carriers by the SOA prior to entering the fibre transmission link that was exposed to varying temperature (20 °C to 50 °C) will influence the recovered OCDMA auto-correlation and its width at the receiver site. We have shown both experimentally and by using simulations that the distorted OCDMA auto-correlation due to temperature induced fibre dispersion ($D_T < 0$) can be corrected by manipulating the chirp of OCDMA code carriers when traversing the biased SOA before entering the transmission link. In addition, the effects of varying values of GVD (parameter β_2) on the OCDMA auto-correlation width were also investigated. We have shown that the impact of a changing β_2 (0.03–0.06) ps^2/nm when the fibre link was exposed to 50 °C could be compensated by controlling the code carriers' chirp by placing an SOA on the transmitting site. Our simulation results are in very good agreement with experimental observations.

6. Conclusions

The real communication systems that were operating under changing environmental conditions may experience temperature variations along the transmission link. These variations may not be uniform and would locally change the transmission link properties. The local temperature changes will inflict local GVD changes, which in turn will cause deviations from the 'originally designed steady state' fibre dispersion compensation. Since the dispersion is a cumulative effect, it will manifest itself at the receiving end as a time and temperature varying disturbance. To mitigate it would require to dynamically control the SOA.

Author Contributions: I.G. conceived and designed the experiments; contributed to data analyses and writing the paper; supervised the research work; Md.S.A. performed the experiments; contributed to data analyses; designed the analysing tools; contributed to writing the paper.

Acknowledgments: This work was supported by the European Union's Horizon 2020 Research and Innovation Program through the Marie Sklodowska-Curie under Grant 734331.

References

1. Willner, A.E.; Hoanca, B. Fixed and tunable management of fiber chromatic dispersion. In *Optical Fiber Telecommunications IV-B*, 4th ed.; Elsevier: New York, NY, USA, 2002; pp. 642–724.
2. Norgard, G. Chromatic Dispersion Compensation: Extending Your Reach. Available online: http://www.datacenterjournal.com/chromatic-dispersion-compensation-extending-reach/ (accessed on 25 March 2018).
3. Singh, M. Different dispersion compensation techniques in fiber optic communication system: A survey. *Int. J. Adv. Res. Electron. Commun. Eng.* **2015**, *4*, 2236–2240.
4. Karar, A.S.; Cartledge, J.C.; Harley, J.; Roberts, K. Electronic pre-compensation for a 10.7-Gb/s system employing a directly modulated laser. *J. Lightwave Technol.* **2011**, *29*, 2069–2076. [CrossRef]

5. Yang, G.C.; Kwong, W.C. *Prime Codes with Applications to CDMA Optical and Wireless Networks*; Artech House: Boston, MA, USA, 2002.

6. Tanceski, L.; Andonovic, I. Wavelength hopping/time spreading code division multiple access systems. *Electron. Lett.* **1994**, *30*, 1388–1390. [CrossRef]

7. Yegnanarayanan, S.; Bhushan, A.S.; Jalali, B. Fast wavelength-hopping time-spreading encoding/decoding for optical CDMA. *IEEE Photon. Technol. Lett.* **2000**, *12*, 573–575. [CrossRef]

8. Glesk, I.; Prucnal, P.R.; Andonovic, I. Incoherent ultrafast OCDMA receiver design with 2 ps all-optical time gate to suppress multiple-access interference. *IEEE J. Sel. Top. Quantum Electron.* **2008**, *14*, 861–867. [CrossRef]

9. Dang, N.T.; Pham, A.T.; Cheng, Z. Impact of GVD on the performance of 2-D WH/TS OCDMA systems using heterodyne detection receiver. *IEICE Trans. Fundam. Electron. Commun. Comput. Sci.* **2009**, *94*, 1182–1191. [CrossRef]

10. Ng, E.K.; Weichenberg, G.E.; Sargent, E.H. Dispersion in multiwavelength optical code-division multiple-access systems: Impact and remedies. *IEEE Trans. Commun.* **2002**, *50*, 1811–1816. [CrossRef]

11. Idris, S.K.; Osadola, T.B.; Glesk, I. Investigation of all-optical switching OCDMA testbed under the influence of chromatic dispersion and timing jitter. *J. Eng. Technol.* **2013**, *4*, 51–65.

12. Eggleton, B.J.; Ahuja, A.; Westbrook, P.S.; Rogers, J.A.; Kuo, P.T.; Nielsen, N.; Mikkelsen, B. Integrated tunable fiber gratings for dispersion management in high-bit rate systems. *J. Lightwave Technol.* **2000**, *18*, 1418–1432. [CrossRef]

13. Tanizawa, K.; Hirose, A. Adaptive control of tunable dispersion compensator that minimizes time-domain waveform error by steepest descent method. *IEEE Photon. Technol. Lett.* **2006**, *18*, 1466–1468. [CrossRef]

14. Sano, A.; Kataoka, T.; Tomizawa, M.; Hagimoto, K.; Sato, K.; Wakita, K.; Kato, K. Automatic dispersion equalization by monitoring extracted-clock power level in a 40-Gbit/s, 200-km transmission line. In Proceedings of the 22nd European Conference on Optical Communication, Oslo, Norway, 19 September 1996; Volume 2, pp. 207–210.

15. Ooi, H.; Nakamura, K.; Akiyama, Y.; Takahara, T.; Terahara, T.; Kawahata, Y.; Isono, H.; Ishikawa, G. 40-Gb/s WDM transmission with virtually imaged phased array (VIPA) variable dispersion compensators. *J. Lightwave Technol.* **2002**, *20*, 2196–2203.

16. Neilson, D.T. Advanced MEMS devices for channelized dispersion compensation. In Proceedings of the Optical Fiber Communications Conference (OFC2004), Los Angeles, CA, USA, 22 February 2004; Volume 1.

17. Wielandy, S.; Westbrook, P.S.; Fishteyn, M.; Reyes, P.; Schairer, W.; Rohde, H.; Lehmann, G. Demonstration of automatic dispersion control for 160 Gbit/s transmission over 275 km of deployed fibre. *Electron. Lett.* **2004**, *40*, 690–691. [CrossRef]

18. Ahmed, M.S.; Abuhelala, M.S.K.; Glesk, I. Managing dispersion-affected OCDMA auto-correlation based on PS multiwavelength code carriers using SOA. *IEEE/OSA J. Opt. Commun. Netw.* **2017**, *9*, 693–698. [CrossRef]

19. Ahmed, M.S.; Glesk, I. Mitigation of temperature induced dispersion in optical fiber on OCDMA auto-correlation. *IEEE Photon. Technol. Lett.* **2017**, *29*, 1979–1982. [CrossRef]

20. Ahmed, M.S.; Glesk, I. Management of OCDMA auto-correlation width by chirp manipulation using SOA. *IEEE Photon. Technol. Lett.* **2018**, *30*, 785–788. [CrossRef]

21. Huang, J.F.; Yang, C.C. Reductions of multiple-access interference in fiber-grating-based optical CDMA network. *IEEE Trans. Commun.* **2002**, *50*, 1680–1687. [CrossRef]

22. Kato, T.; Koyano, Y.; Nishimura, M. Temperature dependence of chromatic dispersion in various types of optical fiber. *Opt. Lett.* **2000**, *25*, 1156–1158. [CrossRef] [PubMed]

23. Agrawal, G.P. *Nonlinear Fiber Optics*; Academic Press: Cambridge, MA, USA, 2007.

24. Osadola, T.B.; Idris, S.K.; Glesk, I.; Kwong, W.C. Effect of variations in environmental temperature on 2D-WH/TS OCDMA code performance. *J. Opt. Commun. Netw.* **2013**, *5*, 68–73. [CrossRef]

25. Wright, E.M. Module 3–Numerical Pulse Propagation in Fibers. Available online: http://data.cian-erc.org/supercourse/Graduatelevel/module_3/3_SC_GRAD_LEVEL_Module3_wright.pdf (accessed on 25 March 2018).

26. Abramczyk, H. Dispersion Phenomena in Optical Fibers. Virtual European University on Lasers. Available online: http://www.mitr.p.lodz.pl/evu/wyklady/ (accessed on 25 March 2018).

27. Tsai, C.Y.; Yang, G.C.; Lin, J.S.; Chang, C.Y.; Glesk, I.; Kwong, W.C. Pulse-power-detection analysis of incoherent O-CDMA systems under the influence of fiber temperature fluctuations. *J. Lightwave Technol.* **2017**, *35*, 2366–2379. [CrossRef]
28. Majumder, S.P.; Azhari, A.; Abbou, F.M. Impact of fiber chromatic dispersion on the BER performance of an optical CDMA IM/DD transmission system. *IEEE Photon. Technol. Lett.* **2005**, *17*, 1340–1342. [CrossRef]
29. Islam, M.J.; Halder, K.K.; Islam, M.R. Effect of optical pulse shape on the performance of OCDMA in presence of GVD and pulse linear chirp. *Int. J. Comput. Sci. Eng.* **2010**, *2*, 1041–1046.

MDPI

St. Alban-Anlage 66

4052 Basel

Switzerland

Tel. +41 61 683 77 34

Fax +41 61 302 89 18

www.mdpi.com

Applied Sciences Editorial Office

E-mail: applsci@mdpi.com

www.mdpi.com/journal/applsci

www.ingramcontent.com/pod-product-compliance
Lightning Source LLC
Chambersburg PA
CBHW041218220326
41597CB00033BA/6034